DATE DUE

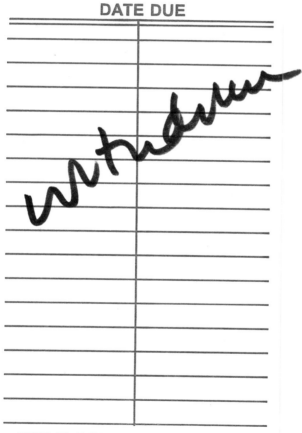

DEMCO, INC. 38-2931

Planetary Dreams

Books by Robert Shapiro

Origins: A Skeptic's Guide to the Creation of Life on Earth

Life Beyond Earth: The Intelligent Earthling's
Guide to Life in the Universe
(with Gerald Feinberg)

The Human Blueprint: The Race to Unlock
the Secrets of Our Genetic Script

Drawing by William R. Leigh from H. G. Wells, "The Things That Live on Mars," a nonfiction article that appeared in *Cosmopolitan*, March 1908.

Planetary Dreams

The Quest to Discover Life beyond Earth

Robert Shapiro

John Wiley & Sons, Inc.

New York • Chichester • Weinheim • Brisbane • Singapore • Toronto

Library of Congress Cataloging-in-Publication Data:

Shapiro, Robert   *12903*
    Planetary Dreams : the quest to discover life beyond earth / Robert Shapiro.
        p.   cm.
    Includes index.
    ISBN 0-471-17936-1 (cloth : alk. paper)
        1. Life on other planets.   2. Extraterrestrial anthropology.
    I. Title.
QB54.S46   1999   *576.839*
576.8'39—dc21   *SHA*   98-35326

Printed in the United States of America

10   9   8   7   6   5   4   3   2   1

Contents

Preface

The ongoing, though intermittent, search for life in our solar system represents the central, unifying theme of this book. A number of researchers believe that a discovery of this type would be one of the most important ever made in science. I agree with them. But other scientists believe that such a search represents an enormous waste of public money. They feel that only Earth, in this system, provides a suitable home for life. Some of them argue that Earth may hold the only intelligent life in the entire universe.

Many of our fellow citizens may also see little point in a search for life in the solar system, but for very different reasons. They believe that intelligent aliens have been orbiting our planet for some time, occasionally descending to abduct suitable human specimens for genetic experiments. If advanced extraterrestrials are already here, then why search for much less evolved forms on Mars?

Such quarrels are not new. The debate over extraterrestrial life has been carried out with a great deal of passion, but with little progress, for centuries. Only in the last decades have we gained the ability to move it forward by collecting data at close range. We can send robots to inspect likely worlds such as Mars, Europa, and Titan, and return photographs, information, and samples, or, if we choose, we can go there ourselves and look around. We may find existing life, remnants of extinct life, or chemical systems evolving in the direction of life. Alternatively, we may encounter monotonous wastelands, lacking any sign that a process relevant to life has taken place there.

The results will help decide which of two very different views of the universe is more nearly correct. In one, the universe, despite its size, is barren. Life started on this planet either through special divine intervention or, for the nonreligious, by an enormous once-in-a-universe stroke of luck. The other point of view holds that the universe is fertile. The circumstances that permit life are inherent within the laws of nature. Life will begin naturally and in a variety of circumstances, once certain basic conditions have been satisfied. If we only look, then we will discover a cosmos rich in life.

This debate cannot be settled by writing a book, but we may find the answer if we inspect the most likely sites in our solar system. I have tried to explain why certain sites seem very promising from a biochemist's point of view, and what we may find when we explore them. Our encounter with reality will culminate a long process in which human beings have projected their desires and fantasies onto our neighboring worlds. For millennia, scientists and others who are fascinated by the planets have conjured up exotic images of the beings that may dwell there. Many of them were banished as we gained a deeper understanding about what our neighboring worlds were really like. Yet the central idea, that life is not confined only to the planet Earth but has sprung up elsewhere in this vast universe, remains possible, though hotly disputed.

To appreciate the magnitude of the past debate and understand the feelings that have carried over into the present time, we will tour some of those earlier visions, the "dreams" that I refer to in my title. Our heritage of imagined fact and admitted science fiction concerning the planets remains rich and entertaining, and I have not resisted the temptation to embellish it a bit in this book.

The "dreams" that I write of are not the usual ones, the images that come up in our minds involuntarily during certain stages of sleep, but rather the hopes and expectations that we have lavished upon the other worlds around us. The other term in the title of this book also has an unconventional meaning. I will use the term *planets* to describe all of those worlds large enough to capture our imaginations. Strictly speaking, the term should only be applied to large bodies that circle the Sun directly, and not to the satellites or moons that orbit the planets. I felt, however, that a single evocative word should be used to include all of the larger worlds that we will want to explore, and perhaps inhabit, in the future.

In examining the planetary dreams of the past, I have tried to connect them to our picture of the larger universe that surrounded these worlds, and what role this larger cosmos played in the unfolding story of human history. Problems arise when we attempt to describe our present view of the dimensions and age of the universe in a way that does not overwhelm the reader. I have attempted to humanize the process by describing it in terms of a visit to a museum, an excursion which we can experience in our present lives. Yet we must keep some measuring stick at hand so that the comparisons remain meaningful. I have chosen to work only with the metric system in terms of length, as I felt that the continual insertion of equivalents in terms of miles and inches would clutter the text. For readers unfamiliar with metric units, the following conversions will help: A kilometer is about six-tenths of a mile, but little harm will be done here if you simply divide the number of kilometers by two, to convert it to miles. In the same spirit, a meter may be taken as a yard, though it is about 10 percent longer. A centimeter is a bit more troublesome—two and a half of them make an inch. Questions of weight and temperature come up much less frequently in this book, so I have included both the metric units and the more familiar terms used in the United States side by side.

I hope that this book can be read simply for entertainment, or for the information that it contains, but I do have an additional agenda in mind. A generation ago, I and many of my friends were very excited by our exploration of the Moon and our robot reconnaissance of other nearby worlds. We have been dismayed to see this outburst of energy dissipate and dwindle to a trickle of missions of interest only to scientific specialists. The question of extraterrestrial life has had a very minor place in the recent robotic explorations. By emphasizing the importance of this question, I hope to move it to the front of the agenda and to infuse new energy into the space program as a whole. I could not get into this topic, however, without bringing up the other compelling motive for a larger human presence in space: The long-term survival and prosperity of the human race depends on it. I feel that these two purposes are linked psychologically and can move together side by side. If this work can play some role in accelerating that movement, then I shall be very gratified.

Robert Shapiro
New York City
June 1998

Acknowledgments

I dedicate this book to the memory of my lifelong friends, Gary and Menasha, I miss them very much.

My interest in this area began many years ago when I was approached by my close friend, Gerald (Gary) Feinberg, to collaborate in an earlier work, *Life Beyond Earth*. Many of the concepts that I consider here originated in our discussions for that earlier work. Our continuing conversations, until the time of his death, helped me to develop my ideas further. I deeply regret that he cannot share his thoughts about the final work with me.

Many scientists helped me during the preparation of this book by sharing their expertise with me in face-to-face discussions or interviews or by electronic mail. They include Mark Adler, Gustaf Arrhenius, John Baross, Daniel Britt, Graham Cairns-Smith, Sherwood Chang, Julian Chela-Flores, Brian Cooper, John Cronin, Christian De Duve, John Delaney, Albert Eschenmoser, Jack Farmer, Jim Ferris, E. Imre Friedman, J. Richard Gott III, Mayo Greenberg, Ralph Greenberg, Hyman Hartmann, Richard Hoover, Joan Horvath, Bob Jastrow, Torrence Johnson, Doron Lancet, Gilbert Levin, Alexandra MacDermott, Cliff Matthews, Chris McKay, Stanley Miller, Bruce Murray, Leslie Orgel, Bill Schopf, Linda Spilker, David Usher, and Gunter Wächtershäuser. Dr. Duane Gish was generous with his time and arranged a guided tour of the Museum of Creation and Earth History for me.

Alice Adler, Victor Baker, Bill Burrows, Ken Edgett, Mike Gentry, Bruce Murray, Jurrie van der Woude, Adrienne Wasserman, and Arthur Winfree provided useful technical materials. Diane Ainsworth and George F. Alexander helped arrange my visit to Jet Propulsion Laboratories at a very busy time.

The following individuals deserve special thanks for reading portions of this manuscript and providing comments: Sallie Baliunas, John Cronin,

Steven J. Dick, John Kerridge, Doron Lancet, Jonathan Lunine, David Morrison, and Daniel Segré.

Bill Burrows, my friend and New York University colleague, provided valuable advice and encouragement at several key stages of the book preparation process. I am grateful for his support.

I am indebted to my agent, Katinka Matson, and her staff for their role in making this book possible. My editor, Emily Loose, provided valuable encouragement and advice, and Diane Aronson and the staff at John Wiley furnished the necessary support and technical assistance in getting this book ready for publication. Patricia M. Daly was a diligent copy editor.

The concentration and diversion of time required for the preparation of a book put special stresses on those who have the misfortune of living with the aspiring author. My wife, Sandy, put up with my moods and provided love and comfort in exchange. She provided vital, if nontechnical, support for this undertaking.

1

Planetary Dreams

Dream not of other worlds; what creatures there
live in what state, condition or degree.
—*John Milton*

The astronaut in his bulky suit moved through the open hatch of the lunar module and onto a large platform. He prepared himself to climb down the ladder that led to the surface. Before he reached the bottom, he pulled a ring on the side of the module to activate a small TV camera. When a message came back that the TV picture was being received, he started down the ladder toward the bottom rung.

A few hours earlier, I had huddled nervously by my radio until the exhilarating words came through: "Houston, Tranquility Base here. The Eagle has landed." Stunned by what had happened, I walked out of my apartment into a summer day in Washington Square Park, New York City. I was surprised to see the arch in its usual place and the crowds going about their normal business. At some deep level, I had expected the world to be rearranged.

That evening, I nailed my gaze to the small television screen and watched with amazement as the first pasty black-and-white images came in from the Moon. We all know what happened next. The astronauts explored the lunar surface for some hours, took photographs, collected rock samples, planted a flag, and safely started their return voyage on the next day.

The event is still recorded in bronze near the main desk of my town

1

library. A copy of the front page of the *New York Times* of July 21, 1969, proclaims in enormous letters: "Men Walk on Moon." Three other plaques complete the display. The librarians had selected replicas of the Genesis page of the Gutenberg Bible, the title page from a 1623 First Folio Edition of William Shakespeare's plays, and the Declaration of Independence. Obviously, great consequences were expected from the Moon excursion.

But events took a very different direction, despite the headlines and bronze plaques. After several more expeditions, the remainder of the Apollo Program was canceled. Human exploration of the Moon ended in 1972. Humans have not returned since then. In the words of Buzz Aldrin, the man who waited in the lunar module as Neil Armstrong climbed down that ladder, "The promise of a sustained, vibrant and growing human presence on the Moon has died a pathetic, almost incomprehensible death."

Yet the reasons for the program's downfall had existed from its very start. They were illustrated in a recent cable television production, *From the Earth to the Moon*. An early meeting of presidential advisers at the White House is restaged, and a science consultant comments, "The only thing we'll get for our money is some rocks."

The Moon, of course, has been obvious in the sky for the entire duration of human history. Men and women have looked up to it and created many different dreams about its meaning. Many of these visions have waned over the centuries, while others kept their substance up to the time of *Apollo 11*. When Armstrong and Aldrin stepped onto the Moon's surface, however, these myths met their doom. A stark reality had replaced them. The true significance of that occasion did not lie in what happened, but what did *not* happen. To make this clear, I will present three accounts of events that definitely did not take place during the *Apollo 11* landing.

A Slip of the Tongue: Non-Happening #1

Neil Armstrong stood on the ladder, glanced down at the ground below, and announced, "I'm going to step off the LM now." He lowered his left foot onto the lunar dust and spoke the prepared words: "That's one small step for man, one giant leap for mankind."

"*Cut*," yelled the director. "Neil, you blew it. Turn the camera off." Armstrong moved the switch on the TV camera, shutting it down, and then walked over to the director. As he walked, he carefully limited his steps to an apron that lay hidden from view of the camera. A worker used the same apron to move onto the set, and removed the single footprint in sand with a brush, as the director explained the blunder: "Neil, the words were: 'That's one small step for *a* man, one giant leap for mankind.' The way you said it," he continued, "makes no sense at all."

"But wait," his assistant said, "I like it better this way. It sounds more natural—the way it really might have happened. Let's check it out with Arthur."

When Arthur agreed with the assistant, Armstrong remounted the module and repeated the scene exactly as before. The remainder of the session went smoothly, and the film could be shown with full confidence when it premiered on July 20, 1969.

Author's Explanation: A Slip of the Tongue

This idea will sound absurd to most of us, but a few individuals have insisted that the Moon landing was a hoax. Perhaps the most prominent was Charles Johnson, President of the Flat Earth Society. In a 1980 interview, he maintained that the space program existed mainly to prop up the myth that the Earth is a globe. "The known, inhabited world is flat," he insisted. "Just as a guess, I'd say that the dome of heaven is about 4,000 miles away, and the stars are about as far as San Francisco is from Boston." He concluded, "The Sun and Moon are about 3,000 miles away and 32 miles across."

According to Johnson, the Moon landings were faked by Hollywood studios. The noted science fiction writer, Arthur C. Clarke, wrote the scripts. Johnson continued, "I recommend that the government get out of the space business and turn the whole thing over to ABC, CBS and NBC. The TV networks do a far superior job."

The source of his information was not simply the interior of his cranium. In 1968, one year prior to the first Moon landing, the stunning science fiction film, *2001, A Space Odyssey*, was released. Arthur C. Clarke had written a science fiction novel with the same name and had coauthored the screenplay with producer-director Stanley Kubrick. Their effort won an Oscar for Special Effects. Early in the film we saw realistic renditions of space travel to the Moon, an elaborate space station, and astronauts standing on a lunar hill, just above their well-developed main base. The landing of a shuttle at the base was depicted in exquisite detail, as was the excursion of a lunar flier that visited a site where an important discovery had been made. Considerable pains were taken to include the minor aspects of life in space. (I still recall the instruction sheet that accompanied the space toilet.)

By comparison, the *Apollo 11* transmission looked like an amateur home movie, in washed-out and flickering black and white. The landscape appeared rounded and dull, with none of the dramatic sharp mountain peaks and jagged valleys shown in *2001*. *Apollo 11* had exactly the appearance that one would expect if the production were placed in the hands of government bureaucrats. They would skimp on film quality, even though the real savings would come from using facades instead of rockets and carrying out only a make-believe space program, instead of the real thing.

Charles Johnson's belief system had far deeper roots than distrust of the U.S. government, however. He was a Christian fundamentalist who relied on the Bible for his picture of the universe. He felt that the biblical view limited humankind to its natural domain—a flat Earth, while the heavens were reserved for God and his angels. For example, consider the following quote from Psalms 115:15–16 of the King James version: "Ye are blessed of the LORD which made heaven and earth. The heaven, *even* the heavens, *are* the LORD's; but the earth has he given to the children of men."

Physicist Harold Morowitz has summarized this biblically inspired universe:

> The Book of Genesis is explicit about the earth's being at the hub, leaving all other celestial objects in an accessory role.
>
> To believers in the exact word of Genesis, man is the raison d'être of creation, and all the rest of the universe consists of a group of heavenly objects set there by the Creator to decorate the human abode.

The Bible was not isolated in this interpretation: Other sources from antiquity held similar views and provided additional details. Aristotle, for example, had written, "there must be only one center and circumference; and given this latter fact, it follows from the same evidence and by the same compulsion, that the world must be unique. There cannot be several worlds."

In Aristotle's scheme, the Sun, Moon, planets, and stars were objects set in concentric spheres, fifty-six in number, which revolved around the Earth at the center. The system was perfected by Claudius Ptolemy, a Greek scientist of the second century A.D., and was later adopted by the medieval Roman Catholic Church as part of its unified worldview. If one accepts the

idea, dominant in our culture for many centuries, that the planets are dec-
orations on a celestial ceiling overhanging our Earth, then any thought of
pedestrian traffic on their surface appears nonsensical.

When Armstrong stepped down on the Moon's surface, he ended sym-
bolically a centuries-long process in which the the Copernican sun-at-the-
center system was substituted for the one of Ptolemy. The changeover did
not take place easily. Galileo Galilei (1564–1642) was the first to observe a
number of heavenly bodies through a telescope. He concluded that the
Moon and planets were worlds in their own right. As a result, he came to
no small measure of grief. He was compelled by the Catholic Church to reject his support for the Copernican system. Italian philosopher Giordano
Bruno (1548–1600) had fared worse. He was burned at the stake in Rome,
in part because of his advocacy of a multitude of worlds. Only within the
past few years has Galileo been fully absolved from his heresy.

By the twentieth century, however, virtually everyone in technological
societies accepted that the planets were separate worlds. Yet some final ceremony was appropriate. When Neil Armstrong and Buzz Aldrin actually
walked on the surface of another world, with a sky above, ground beneath
their feet, and Earth reduced visually to an ornament in the heavens above,
only then was this ancient quarrel appropriately set to rest. To mark the occasion, Pope Paul VI at the Vatican observatory greeted the "conquerers of
the moon, pale lamp of our nights and our dreams." The place of the
Moon is no longer an issue, even in religion—except, of course, for Mr.
Johnson and a handful of others who believe that the U.S. government is
capable of simulating or concealing almost anything, the heavens included.

A View of the Crater: Non-Happening #2

Neil Armstrong moved awkwardly across the gray lava field around the
module, which was rougher than had been expected. He had planted the
flag and collected his rock sample—now he needed to find a suitable place
to relocate the television camera, so that the Earth-bound audience would
have a prettier view. He noticed that the ground rose up behind the lunar
module, and he considered the possibilities of that location. When he had
first tried to land the module, Armstrong had come upon a hazardous
crater at the last minute. To avoid it and find a more suitable flat location,
he had extended the flight, almost exhausting the available fuel. As he flew

over the crater, his view had been obscured by a haze, which he believed was caused by dust from the rockets.

Now he hoped that the rise represented the lip of that crater. By placing the camera there, he could send panoramas of both the landing site and the crater back home. The module photographed well, but he noticed that the haze had persisted in the crater, long after the dust should have settled in a vacuum. In spite of everything the astronomers had predicted, it appeared that some pockets of an atmosphere existed on the Moon. Armstrong attached a set of specially constructed lunar field glasses to his visor and looked more deeply into the crater, then gasped at what he saw and blurted out into the spacesuit's communicator: "Houston, you won't believe this!"

The crater below widened into a meadow and then a valley. The meadow appeared to contain a green, grasslike substance, upon which a group of four-legged, bisonlike creatures grazed. The creatures were novel in that they had a fleshy, hairy appendage that extended across their foreheads from ear to ear. They raised and lowered this flap by using their ears as they moved about, probably as a way to protect their eyes from the lunar extremes of shadow and light. The lunar bisons were kept company by some other creatures that resembled unicorns and some cranelike birds. Farther down the crater, pigmy zebras, long-horned giraffes, and common sheep could be observed nestling among a grove of lunar pine trees.

As Armstrong watched in amazement, four flocks of large winged creatures took off from the far walls of the crater, descended slowly to the plain, and landed near what appeared to be an elaborately constructed temple. He reported all this as best he could to Mission Control Center and received in return an inquiry as to the state of his mental health, together with a request that Buzz Aldrin confirm the information. Armstrong was then joined by Aldrin, who gave a similar account. Working together, they managed to improve the television transmission, which finally showed a blurred view of animals apparently moving on the surface of the Moon.

At this point, the picture transmission from the Moon was suddenly discontinued, and a bland test pattern took its place. However, a heated discussion in the Houston Control Room could still be heard on the audio. This also stopped after a few moments, and an announcement was made that coverage of the Moon landing had been interrupted due to a technical malfunction but would be restored as soon as possible. Private communication between the astronauts and Houston continued, however. Armstrong reported that the winged creatures had taken off again and were coming closer. He was now able to make out much more about their appearance. They were humanoid rather than birdlike. They were about four feet in height and had hairless wings that extended from their shoulders to their legs. Short, glossy, copper-colored hair covered their entire bodies, except in the area of their yellowish faces, which were bearded, with a prominent forehead and humanlike lips. They were definitely headed in the astronauts' direction.

Armstrong urgently requested instructions, as this situation had not been covered in the flight plan. Houston replied: "Stall them awhile if you can. See if they are friendly, but don't commit us to anything. We are setting up hot lines to the Pentagon, the Department of State, and the National Council of Churches!"

Author's Explanation: A View of the Crater

This account of lunar creatures may appear quite ridiculous today and probably would not pass muster with a pulp science fiction magazine. Yet it was taken very seriously in the nineteenth century. The descriptions of animals and bat-men on the Moon are not mine. I have adapted them from a front page news article that appeared in the *New York Sun* of Tuesday, August 25, 1835, and the continuations published by the *Sun* for the remainder of the week. The "observations" were not made by astronauts, of course, but by Earth-based scientists who used a revolutionary giant telescope. The instrument had been constructed near Capetown, South Africa, by the noted British astronomer Sir John Herschel.

The articles provided a full account of the financing, design, and construction of the telescope, as well as the building that housed it. The telescope provided a lunar view comparable to that of an unaided human eye at a distance of a hundred yards, with a resolution of one foot. The Royal Society was acknowledged for its support of 10,000 pounds and the King for 70,000 pounds. Thirty-six oxen were needed to pull the 14,826-pound instrument up to its site, and over four days were required for the trip.

The *New York Sun* quoted a supplement to the *Edinburgh Journal of Science* as the source for its revelations, of which my extracts have provided only a modest sample. According to the stories in the *Sun*, Sir John had also observed many other species on the Moon, fine details of the rings of Saturn, and planets circling other suns!

Although no large-type headlines were used to promote the series, the initial issue did claim that "the portion which we publish today is introduction to celestial discoveries of higher and more universal interest than any, in any science yet known to the human race." As we might imagine, it attracted considerable attention.

By August 31, the circulation of the *Sun* had soared to 19,500, the largest on Earth for a newspaper of its type. Unfortunately, the promised supporting documents failed to materialize and a journalist, Richard Adams Locke, then admitted that he had invented the entire episode. A kernel of truth existed, though: Sir John Hershell was in South Africa at the cited time, performing valuable but less spectacular measurements with a much more modest telescope. The event has been celebrated as "The Great Moon Hoax," though historian Michael Crowe has suggested that it was intended as a satire rather than a hoax. As we shall see, there was much to satirize.

Before the truth came out, however, there was apparently little skepticism about the stories, despite their sensational nature. The poet Edgar Allan Poe noted subsequently, "Not one person in ten discredited [the articles]. A grave professor of mathematics at a Virginia college told me that he had *no doubt* of the entire affair." Others claimed that at Yale "nobody expressed or entertained a doubt as to the truth of the story." Another report of the time claimed that a group of clergymen wrote to Herschel "beseeching him to inform [them] whether science affords any prospects of . . . conveying the Gospel to residents of the moon." The *New Yorker* described the discoveries as being "of outstanding merit, creating a new era in astronomy," and the *New York Times* called them "probable and possible."

From today's point of view we may consider such gullibility surprising. But the ideas had not just emerged fresh from Locke's imagination. A great precedent for such revelations had been established in advance in the public mind. Tales of lunar life had been advanced for millennia by scholars of the highest reputation.

The early philosophers who recognized that the planets were separate worlds found it reasonable to assume that they were simply extensions of the one we knew. Theories of this type appeared in Greece in the sixth century B.C., when Thales of Miletus argued that the Moon was similar to the Earth. In 450 B.C. Democritus speculated that the Moon had mountains and valleys, and the influential pythagorean Philothalus (fifth century B.C.) held that the Moon was inhabited. The following quote from a later Greek source known as "pseudo-Plutarch" was provided by historian Steven Dick:

"The moon is terraneous, is inhabited as our earth is, and contains animals of a larger size and plants of a rarer beauty than our globe affords. The animals in their virtue and energy are fifteen degrees superior to ours, emit nothing excrementitious, and the days are fifteen times longer."

Other cultures created their own variations. In the ancient Hindu and Persian traditions, the Moon was the first port of call for souls who had departed recently from Earth. The lunar crescent is swollen in the early part of the month by their arrival, and the full moon wanes as they depart for another destination.

Those stories were constructed from imagination and reason alone, but information was welcome, too, as long as it supported the basic themes. When Galileo turned the newly invented telescope toward the Moon in the early seventeenth century, he observed valleys and mountains; a large circular crater was included in a sketch that he made of the Moon. Astronomer Johannes Kepler had commented in about 1595, "The body of the Moon is of such a kind as is this our earth, comprising one globe out of water and land." When Kepler examined Galileo's sketch, he concluded that the crater "was formed by intelligent inhabitants" who "make their homes in numerous caves hewn out of that circular embankment." He used these inhabitants in a novel that we will mention shortly.

Not all such claims were presented as fiction. In 1638, the Reverend John Wilkins (who was later master of Trinity College, Cambridge, and bishop of Chester) wrote a work entitled "The Discovery of a World in the Moone, or a Discourse Tending to Prove That T'is Probable There May Be Another Habitable World in That Planet," in which he advocated lunar life. According to historian Michael J. Crowe, this was considered the most influential of the works of popular astronomy in that century. "Especially after Wilkins' book, the Moon became the battleground for a war, lasting hundreds of years, concerning lunar life."

The noted astronomer Sir William Herschel (1738–1822), father of Sir John Herschel and discoverer of the planet Uranus, had a large role in that combat. In May 1780, he published a paper entitled "Astronomical Observations Relating to the Mountains of the Moon" in which he claimed that "the knowledge of the construction of the Moon leads us insensibly to several consequences . . . such as the great probability, not to say almost absolute certainty, of her being inhabited." He believed that he had observed forests, cities, canals, roads, and two small pyramids on the Moon.

Herschel was not alone in advancing such concepts. The astronomer and M.D. Franz von Paula Gruithausen was the author of 177 astronomy papers and editor of three journals. In his 1824 paper entitled "Discovery of Many Distinct Traces of Lunar Inhabitants, Especially of One of Their Collossal Buildings," he claimed to have observed vegetation, animals, and geometric features that included roads, walls, fortifications, and a star-shaped structure he labeled a temple. Two years later, he was appointed professor of astronomy at Munich University.

Sir William Herschel's son, John, who was used as the leading character in the *New York Sun* hoax, also believed in extraterrestrial life. He admitted that the Moon "has no clouds, nor any other indications of an atmosphere" but felt that in the past, water may have been present: "There are regions perfectly level, and apparently of a decided alluvial character."

Unfortunately, as astronomical observations improved, our perceptions of the Moon's Earth-like qualities diminished. By the 1870s, scientific treatises were picturing the Moon as "an airless, waterless, lifeless, unchangeable desert" and the lunar landscape as "a realization of a fearful dream of devastation and lifelessness—not a dream of death . . . but a vision of a world upon which the light of life has never dawned." But others still continued to argue that a thin atmosphere was present in certain locations and that color changes were due to vegetation. Perhaps the last such advocate was William H. Pickering (1858–1938), a prominent Harvard astronomer. He maintained that the Moon had a thin atmosphere, snow or frost on its peaks, and vegetation. Dark spots on the surface were claimed to be swarms of lunar insects in migration. He supported his idea until his death, thirty-one years before the *Apollo 11* landing. The belief that some form of life could be found on the Moon had resisted every telescopic revelation and persisted into the twentieth century. To really appreciate what the place was like, we would have to go there in person.

The Exploration Itch

Many of us, when we hear of new and strange places, develop the desire to visit them. The early telescopic discoveries concerning the Moon and planets coincided with a great Age of Discovery, in which European adventurers set sail in their ships to explore vast areas of Earth that had come to their attention for the first time. They departed with the hope of finding treasures and discovering wonders. They did both, and in the process saw peoples and beasts that seemed strange to them: Aztecs and llamas, aborigines and kangaroos. Such overseas encounters had long been embellished and exaggerated in myth and fiction. According to tradition, a variety of strange creatures would be found when we wandered far from home: mermaids and sea serpents, the giant one-eyed Cyclops of Homer, and the intelligent horses and degenerate humans of Jonathan Swift.

If the era of exploration was to be extended to the Moon, then it would only be necessary to substitute space for the oceans of the earlier tales. Suitable ships for this voyage were lacking before our present era, so magical means were substituted. In Johannes Kepler's posthumous novel *Somnium* (1634), demons transport the hero across the bridge of darkness that links the Earth and Moon during an eclipse. Intelligent inhabitants, of course, greet the hero when he arrives.

In subsequent centuries, other writers used magnetic energy, bird power, balloons, and enormous explosions to propel explorers to the Moon. The tradition lives on today in science fiction epics such as *Star Wars* and *Star Trek*, in which space travel has been extended to travel between the stars. The inconveniently large interstellar distances are shrunk to manageable size by the use of fanciful space warps and drives. Most worlds encountered at the other end of such voyages have Earth-like temperatures, breathable air, and inhabitants and beasts that are simple permutations of the ones we know on Earth.

By the nineteenth century, Jules Verne and other writers were giving serious thought to the possibility of reaching the Moon, and in the twentieth century scientists developed the rockets to make travel to the Moon possible. By 1969, we were ready to go. Unfortunately, by that time, the bat-men and unicorns had evaporated, and only more primitive possibilities remained.

Surprise in the Dust: Non-Happening #3

Buzz Aldrin and Neil Armstrong had found an appropriate location for the television camera. Their next order of business was to plant the American flag, and they were looking for a suitable spot. As they searched, they continued their conversation with Mission Control in Houston:

HOUSTON: "We see the shadow of the LM."

ARMSTRONG: "Roger, the little hill beyond the shadow of the LM is a pair of elongated craters about forty feet long and twenty feet across, and they're probably six feet deep. I'll work my way over there and see if there's a suitable spot."

As Armstrong moved along, he commented on the quality of the lunar dust cover:

ARMSTRONG: "The surface is fine and powdery. I can kick it up loosely with my toe. It does adhere in fine layers, like powdered charcoal, to the sole and sides of my boots. I only go in a small fraction of an inch, maybe an eighth of an inch, but I can see the footprints of my boots and the treads in the fine sandy particles. . . . It's a very soft surface, but here and there I run into a very hard surface."

Buzz Aldrin had picked up the American flag and a hammer in anticipation of the coming ceremony. He now watched as Armstrong hopped to the edge of the crater, took an additional step . . . and disappeared. He rushed to the edge of the depression and noticed immediately that the dust was a darker gray than the material closer to the lander. Armstrong was not in sight, but some of the dust was settling, with a whirlpool motion, into a cavity immediately in front of him: Aldrin transmitted immediately. "Houston, we've got a problem."

Neil Armstrong had sunk into a quicksandlike pocket of very fine lunar material.

Fortunately, the crater was not very deep, and with the use of the flag, a long strap called the lunar equipment conveyer, and some advice from Houston, Aldrin was able to help Armstrong pull himself out. Within an hour Armstrong and Aldrin were back in the module, their excursion temporarily halted to check whether Armstrong or his equipment had been injured. The cabin of the module was repressured with oxygen, and the astronauts removed their suits. As their helmets came off, they noticed a pungent odor. Armstrong compared it to wet ashes in a fireplace and Aldrin to gunpowder after an explosion. Armstrong, with Aldrin's help, checked both himself and the suit for any obvious sign of damage and found none.

"Tranquility base to Houston," Armstrong commented. "I think everything is OK and . . ." His comments were interrupted as he suddenly erupted in a loud sneeze. "Houston, this has come on pretty quickly. I may be catching a cold."

Author's Explanation: Surprise in the Dust

As the reality of the Moon became increasingly apparent in the mid-twentieth century, dreams and speculations about travel to the Moon had to be scaled down. Yet many felt, in advance of the *Apollo 11* landing, that some authentic surprises awaited us. Among the speculations were the possibility of deep and possibly hazardous lunar dust. Another was that the lunar dust would ignite when it was exposed to oxygen in the cabin of the lunar module, *Eagle*. Neither event took place. But perhaps the most troublesome concern was the thought of dormant microorganisms on the Moon.

The director of the Harvard Observatory, Donald H. Menzel, had commented in *National Geographic* magazine in 1958: "A thin layer of dust covers the lunar surface from a depth of several inches to perhaps several feet. One prominent scientist" (not named) "has argued convincingly that the dust cover may be miles in thickness and that the craters themselves are dimples in this layer." In the same article, Menzel emphasized the extreme ruggedness of the Moon. A painting that accompanied the article showed tall jagged mountains rising abruptly over a flat, crater-riddled plain.

Astronomer Robert Jastrow advanced the theme two years later in the *Scientific American:* "The hail of meteorites must have chipped away at the rocks, and the moon must be covered with a layer of rock dust. . . . Some unknown agent distributes the dust over the lunar surface with such uniformity that the small pits and craters are filled in." Jastrow then cited the idea of Cornell astronomer Thomas Gold that cosmic bombardment had placed positive charges on the dust particles. Mutual repulsion then makes them "act like Mexican jumping beans." They hop downslope, propelled by gravity. "Thus the moon's highlands should be scoured clean, and its depressions filled with dust to a considerable depth." Jastrow continued: "If Gold is right, it would be well not to choose the otherwise smooth and inviting surface of a lunar sea for the first landing on the moon. The porous rusk might not be able to bear the modest weight of an exploration vehicle, and so might swallow it up."

The Planetary Society described similar ideas that had been put forward by astronomer Fred Hoyle: "Prior to the first Apollo Moon landings, his minority opinion that the surface of the moon was covered with a thick layer of moondust was taken seriously enough so that contingency plans were discussed for the unlikely eventuality that astronauts, or whole spacecraft, might sink out of sight into dust pools on the Moon's surface."

Hoyle worked his thoughts into his 1967 science fiction story "The Martians": "The dust was really nasty stuff. It climbed all over you, hard to see, if you were unlucky enough to step into it. It climbed all over your equipment, into every crevice more than a few microns in size. . . . Unfortunately, the Moon has a lot of dust, so not many places could be found where the first landing module could be set down."

In practice, the *Apollo 11* mission did choose the lunar Sea of Tranquility for the first landing. The surface supported the landing perfectly well, and no untimely burials took place. But a different hazard related to moondust had been anticipated. Nobel Laureate Joshua Lederberg and Dean B. Cowie had speculated in *Science* magazine in 1958 that the dust on the lunar surface might contain material of biological interest.

Lederberg and Cowie proposed that interplanetary dust might be a site where molecules containing the key element carbon would be made. Substances that contain carbon are called organic compounds, and many of them play a central role in the biochemistry of our own kind of life. In the early days of our solar system, carbon compounds would have descended impartially on both the Earth and the Moon. They may have helped life on Earth get started, but they would have been consumed in the process. On the Moon, however, they may have survived to this day, providing us with a record of our own beginnings. The scientific value of such a discovery would have been enormous, but Lederberg and Cowie suggested that more than organic chemicals might be lurking in the moondust. Microbes might be waiting there as well, in frozen storage.

This possibility arose from a theory called panspermia, which had been publicized early in the twentieth century by Nobel Laureate chemist Svante Arrhenius. Arrhenius had suggested that living microorganisms can travel from planet to planet, and perhaps even between solar systems, by drifting through space in a dormant form. (We shall wade more deeply into this topic in Chapter 7.) If this idea was correct, then such microbes might have drifted onto the Moon and would be waiting in the dust.

Apart from panspermia, another possibility had been raised that would have showered biological materials onto the Moon. Once again, the author was a Nobel Laureate in chemistry, Harold C. Urey. Urey suggested that the Moon approached the Earth so closely during the early days of the solar system that a portion of the ocean and whatever it might have contained were transferred to the lunar surface. So a number of possibilities existed in which our astronauts might have met some of their earlier ancestors on the Moon and perhaps caught cold from them.

Although the idea of forests and animals on the Moon had been dismissed a century ago, the fear of microorganisms there remained very much alive at the time of *Apollo 11*. This threat was taken seriously enough that the astronauts were given biological isolation garments while they were still in the spacecraft, immediately after they landed in the Pacific Ocean. Dressed in these garments, they transferred to a life raft that contained a decontamination solution. The spacecraft's hatch was then closed to prevent the escape of possible lunar microorganisms. Father Eugene Cargill, who was visiting Neil Armstrong's wife, proposed a champagne toast at that point: "Here's to Jan and Neil, whom we'll never be able to visit again because of fear of contamination."

Upon arriving aboard the aircraft carrier *Hornet,* the astronauts were placed in a trailer, which was hermetically sealed. They were then transferred to the Manned Spacecraft Center in Houston, where they were kept in quarantine within a lunar receiving laboratory. Their quarantine continued until three weeks after their liftoff from the Moon. Not every scientist has agreed that these measures were required. Edward Anders of the University of Chicago wrote a letter to the *New York Times* in which he offered to swallow a sample of the lunar dust. His offer was not taken up, but his attitude was proven correct by later events. It was only after the *Apollo 14* mission that the entire quarantine procedure was discarded. The lunar microorganisms had never appeared, and it was recognized that the Moon was stone cold dead.

Apollo 11 was a great extinguisher of dreams. When the astronauts stepped onto the actual surface of the Moon and looked around, the visions of many generations and cultures evaporated. The ideas of Thales, Ptolemy, the Herschels, and Urey could now share the same trash bin. What remained was the stark reality that the Moon imposed on us when we chose to confront it at point blank range. Six manned missions explored the lunar surface, from *Apollo 11,* which landed on July 20, 1969, to *Apollo 17,* which departed on December 13, 1972. They returned with souvenirs of the places they visited: 382 kilograms (840 pounds) of assorted rocks. From their observations and these materials, scientists were able to reconstruct the authentic story of the Moon. In exchange for our dreams, we gained a lunar history.

A Brief History of the Moon

Our nearest neighbor was born under very violent circumstances, about four and a half billion years ago. At that time, the Earth itself was only a few million years old. Our planet was still in the final stages of assembly from the small pieces of debris that abounded in the early solar system when it was hit by another body. This event has been called the "Big Whack," in the tradition of the Big Bang, the explosion that launched the universe. The other party to this massive collision was a planet about a quarter of Earth's mass. Before the smashup the Earth, and perhaps its Whack partner as well, had undergone some internal settling, with their heavy iron components sinking to their centers. After the crash, the iron cores of both bodies merged to form the center of the remodeled Earth, and some of the lighter surface portions were knocked loose into space. Some of the debris returned to Earth, and the rest formed the Moon. This newly born world was less dense than Earth and lacked water and some of the lighter elements.

A lot of energy was released by the collision as well, and as one result, the newly formed Moon was left with a surface of molten rock, which gradually solidified over millions of years. This fresh, virginal surface was quickly pockmarked, however, by the continual infall of smaller pieces of solar system debris. The cratering process occurred intermittently, with a crescendo at about 3.8 to 4.0 billion years ago. Thereafter, it fell off sharply, but it still continues to this day. Most scientists feel that this last heavy cratering episode was the result of a catastrophe that affected the entire inner solar system, our own world included. Recently, that viewpoint has been questioned and the possibility has been raised that only the Moon was affected. The details of this scenario have important implications for theories about how life started on Earth, but we are discussing the Moon right now, so we will postpone this puzzle for later.

Another major event in lunar history took place about 3.3 billion years ago. Internal events (perhaps radioactive heating) within the Moon released heat. Molten rock emerged from the interior and flowed into large depressed areas on the surface that had been previously hollowed out by crater impacts. When this lava had solidified, it appeared darker than the surrounding older highland area. The result is visible to us as the familiar face of the "Man in the Moon." Since then, all has been relatively quiet. The interior has gradually cooled down, and occasionally meteorites have gouged out additional large and small craters. Moonquakes are weak and happen rarely, and with no atmosphere present, neither wind nor rain disturbs the surface. The Moon sits peacefully in midlife, awaiting its destruction billions of years from now when the Sun runs out of fuel.

A Flood of Information

The most important question about the Moon concerned its origin, and it did not yield an answer easily under our personal inspection. Before Apollo, three theories had shared the stage: (1) The Moon, already in a wandering existence, may have been captured by the Earth. (2) Both Earth and Moon may have coalesced from debris, side by side. (3) The Moon may have spun from the early Earth, perhaps leaving a depression the size of the Pacific Ocean.

The Apollo data had undermined all three of these ideas. In the late 1970s some scientists joked that perhaps the Moon was just an error of astronomical observation and didn't really exist. But the stage had been cleared, allowing the theory of a great collision to emerge and gain consensus at a conference in 1984. This theory is still vulnerable and may change further. If it remains in place, it will not greatly affect our outlook on the universe, as it describes an individual event of history and not a general law. One prime use may be to enable some parents to answer the question, "Daddy [or Mommy], where does the Moon come from?"

A second question in the same category could be, "How did the Moon get those dark splotches?" The answer I described previously may not have much emotional impact, but it does put to rest a problem that has intrigued humans over the ages.

In the *Divine Comedy* of the great Italian poet Dante Alighieri (1265–1321), the author, having experienced Hell and Purgatory, is guided through Paradise by the spirit of Beatrice, his deceased ideal love. He asks her to account for the appearance of the Moon. In answering, she first rejects an explanation that the dark areas represent Cain's sacrifice of a thorn bush. She also sets aside the idea that certain parts of the Moon are at a farther distance from us and reflect less light. She then makes an analogy to a work of metal made by a smith. Different strokes of his hammer may produce facets that reflect different qualities of the metal. In the same way, different parts of the Moon reflect the varying virtues of heaven; we perceive these as dark and bright areas.

When Galileo first looked at the Moon in 1609, he discovered that the surface was irregular. The dark areas appeared smooth, and the light ones ragged and mountainous. As a result, Galileo concluded that the dark areas were oceans and light continents. Subsequent observers agreed and called the dark regions "maria," or seas. Thus the *Eagle* of *Apollo 11* landed on Mare Tranquillitatis, the Sea of Tranquility. These seas of the imagination evaporated over the centuries, but the final solution to the question of the dark areas had to wait for human arrival on the Moon.

In addition to these highlights, a great amount of information about the Moon was produced by Apollo. We learned about the age and the elemental and isotope composition of minerals from a variety of locations, the exact distance of the Moon from the Earth, the distribution of craters, the nature of the weak lunar magnetic field, and, by monitoring moonquakes, something about the internal structure of the Moon. But even as scientific data were pouring in, the program was being curtailed drastically. By January 1970, the final planned mission of the series, *Apollo 20*, was cut. In March, the White House devalued the priority of space in national affairs and ruled out any further ambitious manned programs. Finally, on September 2, 1970, the announcement was made that the last two missions then scheduled, *Apollo 18* and *Apollo 19*, had also been canceled. On December 20, 1972, Jack Schmitt and Gene Cernan lifted off from the Moon, marking the end of manned lunar exploration for their century. *Science* magazine subsequently published on its cover a cartoon of one lunar rock confessing to another that it would "miss those Apollo chaps."

The Vision That Vanished

Much more had been expected. Buzz Aldrin later wrote in a letter to the members of the National Space Society that in 1969 he, his "colleagues at NASA, and the millions upon millions of people who watched that grainy black and white television shot that evening felt that the Apollo program was only the beginning . . . that human occupation of the Moon and space flights beyond would follow inexorably and quickly."

So inevitable did this human march to the planets seem that nearly 100,000 individuals booked reservations on Pan Am's planned Moon shuttle, and many expected to make the trip in their lifetimes.

Pan Am's was not the first or only such booking. I remember that as a child, in the 1940s, I put my name on a list collected by the Hayden Planetarium in New York City. The list reserved one seat on the first passenger flight to the Moon, which was assigned a date in the 1990s.

In fact, in the late 1960s, administrator Tom Paine and other planners at NASA had planned a future that included a twelve-person space station and reusable shuttle by 1975. The planners envisioned that by 1980, a fifty-person station would be built. An additional platform in lunar orbit, a permanent Moon base, and a manned expedition to Mars were all foreseen before the end of the century. The Mars mission received the enthusiastic support of Vice President Spiro Agnew. These same ideas had been proposed by space pioneer Wernher von Braun and others in the pages of *Collier's* magazine twenty years earlier. And of course, these developments had been rendered visually by Arthur C. Clarke and Stanley Kubrick in *2001, a Space Odyssey.*

Immediately after the *Apollo 11* landing, Clarke wrote in an epilogue to the book *First on the Moon,*

> Anything written about the moon at the beginning of the 1970's will probably look silly in the 1980's and hilarious in the 1990's—particularly to the increasingly numerous inhabitants of our first extraterrestrial colony. . . .
>
> And well before the end of this century, the first human child will be born there. It would be interesting to know the nationality of its parents; but such fading symbols of the old world will not be long remembered, in the fierce and brilliant light of the lunar dawn.

A vision of this magnitude could not disappear unnoticed. A number of the astronauts in the Apollo Program were stunned and embittered by its abrupt termination and by the end of manned space travel beyond Earth's orbit. To quote Buzz Aldrin once again, "for one crowning moment we were creatures of the cosmic ocean, a moment which may be seen a thousand years hence as the signature of our century. Yet an eerie apathy now seems to afflict the very generations who witnessed that event." Astronaut Stu Roosa, who had seen an unfinished obelisk near Aswan, Egypt, was much more blunt. "I always thought Apollo was our unfinished obelisk. It's like we started building this beautiful thing and then we *quit.* History will not be kind to us because we were *stupid.*"

Some journalists shared Aldrin's and Roosa's sentiments. On the occasion of the twenty-fifth anniversary of *Apollo 11*, John Noble Wilford wrote in the *New York Times*, "But the wonder now evoked by the lunar landing is that the United States so recently had the optimism to conceive such a grand undertaking and the will to see it through in a blink of history's eye—and then so readily abandoned the enterprise. Perhaps that is why an air of melancholy, a dispiriting sense of unfulfilled promise, hangs over the Apollo 11 anniversary, and why the space agency is doing little to celebrate its greatest moment." Frank Rich voiced similar sentiments, in commenting on the film *Apollo 13* for the same newspaper: "The notion that Americans in partnership with their government might accomplish anything in space, let alone on earth, belongs to yesterday. The New Frontier, like the old frontier before it, survives only as a theme park now."

Who Killed Apollo?

The rapid trashing of such high expectations leaves us with this puzzling question: Who killed Apollo? The answer is usually provided in terms of the political climate of Apollo's times, which both ignited the program and quenched it. The Apollo Program was launched during a period when we were in intense Cold War competition with the Soviet Union, and the first manned lunar landing appeared to be a prize in that war. But the pride in our position waned when we got involved in the long war in Vietnam. In the words of journalist Richard Lewis, "The war in southeast Asia, which had been eating into the body politic in America like a cancer, was contributing to the decay of public confidence . . . in the Nixon administration, and in congress. The hostile response to the war policy bred a similar response to other quasi-military programs—and the space effort was regarded by all of its critics as such a program."

In addition to the war, the electorate had other social issues on its mind. Senator William Fulbright, who was a vocal critic of NASA spending, sought to underline his point of view with his comment, "It is a matter of reorienting priorities and sewers are more important than rocks from the Moon."

The Vietnam War is now history, and at this writing, America is in an extended period of prosperity. Social issues are present, as always, but do not have the cutting edge of a generation ago. The sites considered for *Apollo* missions *18, 19,* and *20* were of deep scientific interest and had spectacular locations, such as the giant crater Copernicus. But no one, to my knowledge, has proposed that the project be picked up where we left off. Outside of organizations such as the Planetary Society and the National Space Society, there are few advocates for any human return to the Moon. The reasons for the halt are rooted more deeply.

Was the real cause a failure in communication, in that the scientific value of lunar exploration was never communicated to the public? Journalist Richard Lewis has made this argument: "The scientific achievements of the lunar adventure were reported mainly in the scientific journals, but in such a fragmented and cryptic way that only experts would recognize their significance. The general public, which was financing the lunar ballgame, had little opportunity to learn the score. The consequences were an increase in political hostility toward Apollo and a loss of public interest in it."

Yet others, myself included, feel that the scientific significance was not compelling. In advance of the lunar landings, astronomer Fred Hoyle had predicted that the Moon would be "an uninteresting slag heap." Carl Sagan later wrote that "people . . . for reasons that I believe are fundamentally sound . . . are bored by the Moon. It's a static, airless, waterless, black-sky dead world." This theme has been echoed by Stanford historian Joseph Corn: "People lost interest once the Apollo landings became routine . . . nothing really came of the Apollo program."

Some reactions were more extreme. For example, Henry Gee, an assistant editor of the scientific journal *Nature,* wrote the following in his review of the film *Apollo 13:*

> In that we have not returned, and the sum total of scientific knowledge gained by the Apollo program was not significantly greater than nil, it was an abject failure. NASA's greatest scientific successes have come, as they always have, from unmanned probes. . . . This, then, is the real tragedy of the manned space program, and why the film is so affecting: that so many people on the ground as much as in the spacecraft themselves, should expend so much loyalty, effort and commitment, and put their lives at risk, to an end that was never more than tawdry and is now completely pointless.

Other critics have been angered by efforts to justify the project through a combination of empty publicity and lunar geology. To give one sarcastic example, I can cite Bruce Handy, a senior editor at *Time* magazine. In an article entitled "Fly Me to the Moon," which was published in the *New York Times Magazine*, Handy wrote the following:

> Twenty-five years ago this July 20, astronauts first landed on the lunar surface. They picked up rocks, posed for famous pictures and went home; this was widely regarded as an important step in human evolution. Over the next three years, astronauts returned to the moon five more times, picking up more rocks and posing for less famous pictures, to greater and greater public indifference. Eventually, like any TV show with declining ratings, the moon missions were cancelled and the nation moved on to further evolutionary steps, like learning how to spell General Tso. Today NASA doesn't even have rockets big enough to send men to the moon. . . . But what if there were some really compelling reason why returning to the moon should be a national priority—say there's a natural history museum somewhere that doesn't have a moon rock.

A Drought of Inspiration

Obviously something was missing in the Apollo results, something that would not have been remedied by further landings or better publicity. For a clue, let us turn to Duke historian Alex Roland: "Had we elected to explore the Moon with automated vehicles and invested just a fraction of the cost of Apollo, we would have produced more and better science. There was some good science in the Apollo program. . . . It was not, however, the stuff of myth. Apollo was a human adventure, in which the astronauts took priority over all else. No other myth will work."

I agree fully with this statement. Despite some of the quotes I have provided, the Apollo project did produce an abundance of good science. But it was of the type that would be called "normal science," according to the analysis performed by philosopher Thomas Kuhn in his book *The Structure of Scientific Revolutions*. Kuhn applied this term to specialized work, written in technical terms, that is mostly of interest to other workers in a particular discipline. Most scientific work can be so classified. Of much greater interest are observations that Kuhn called "paradigm-shattering," where "paradigm" was used to describe a broad explanatory concept that unites an entire field.

In the case of the Apollo mission, there was an advance expectation among many of us that something totally unexpected, perhaps even paradigm shattering, would be discovered. When I recently asked a long-term friend (a mathematician, not a geologist) to recall what he might have expected, he could come up with only one concrete idea: Perhaps some unusual mineral, with extraordinary properties, would have formed under the unearthly circumstances of the lunar surface. I had had similar feelings during the *Apollo 17* mission to the Taurus-Littrow area, which had some of the darkest material.

The crew placed at that site included astronaut Gene Cernan and geologist Harrison "Jack" Schmitt, the first Ph.D. scientist to explore the Moon. I remember in particular the moment when Schmitt reported, "Oh hey, there is orange soil." Cernan's comment was, "How can there be orange soil on the Moon?" My own expectation was that something wondrous had been discovered. The astronauts thought that they had discovered an active volcanic vent, which would have been a milestone for lunar geology but had no cosmic implications. Even this expectation failed as later dating showed the material to be 3.7 to 3.8 billion years old. Its fresh appearance resulted from being covered by another lava flow after its formation. The material was unearthed by a meteorite impact much more recently, to await the astronauts' arrival. This enthusiasm, with the disappointing follow-up, reflects the public experience with Apollo.

Good science is always welcome, but the Apollo mission had come at a staggering price, one that would have funded an immense amount of research in most other scientific areas. The scientists involved made the assumption that additional missions would be accepted, at a comparable price. A presidential Space Task Group had convened in 1969 to plan further lunar exploration. Richard Lewis has described the Task Group's conclusions: "A number of geologists believed that a full understanding of the Moon required a full-scale geological survey lasting many years, perhaps a century. More than one base would be required." We have seen the outcome: Rather than undertake such a task, the administration abandoned further exploration.

The last part of the quote from Alex Roland is even more important, however, and I would change it only by replacing his word "myth" with "dreams." The Apollo mission destroyed the previous set of dreams concerning the Moon. But no replacement was suggested by those who ran the project—no longer-range goal or sense of purpose was offered with the achievement. Yet inspirational material surrounded the events on every side. I can supply examples with the greatest of ease.

Item: On the same front page of the *New York Times* that announced "Men Walk on Moon" was a poem by Archibald MacLeish, written for the event. The poem celebrated the occasion on which we humans had reached the unattainable and sifted with our hands the sands of the beaches of the Moon.

Item: Consider astronaut Dave Scott's reaction as he stood on the lunar surface during the *Apollo 15* landing. Under a black sky with a burning sun, scenes were laid out with startling clarity. The Apennines, one of the Moon's large mountain ranges, lay on the horizon. Craters scarred the land; some were worn and others sharp rimmed. A tiny speck was visible in the distance: Scott's module, more than five miles away. Scott and Jim Irwin had used a battery-powered rover to move that distance. Armstrong and Aldrin had never ventured more than sixty meters from their lander. "This view was so breathtaking, Scott said, that for a moment it distracted him from the real reason they were there: to hunt for geologic treasure." But perhaps the view and the experience *were* the treasure.

Item: Recall the photograph of a suited Buzz Aldrin on the Moon, as taken by Neil Armstrong, or the Earth as seen from space, with its blue oceans peeking through the cloud cover. I can call them up easily from memory. Millions or billions of others on this globe can do so as well. Such images have become symbols of our century.

Item: I visited the NASA Air and Space Museum in April 1996. Among a host of attractive exhibits, I was particularly impressed by an area devoted to World War I. In adjacent rooms, I saw a canteen, an aircraft shelter, a biplane, and information on the Red Baron and other aces. Though the exhibits were very well done, these rooms were almost empty. The most crowded exhibit by far, among a multitude of aeronautical treasures, was a model of an astronaut with his rover on the Moon. The crowd, somehow, sensed a deep meaning in that exhibit.

In the remainder of this book, I shall provide my ideas about the new dreams that we may follow in a quest to explore the planets. This presentation will take some time and space. Before we set out, I will let you know where we are headed and briefly describe the ends that we may seek out in space. Before I do, I must concede that we are not all fated to follow the same dreams. Many people believe that the fate of the human species is linked to this planet and that outer space is simply a meaningless distraction. One eloquent statement of this position was made by Kurt Vonnegut in *The Sirens of Titan:*

> What mankind hoped to learn in its outward push was who was actually in charge of all creation and what creation was all about.
>
> Mankind flung its advance agents ever outward, ever outward. Eventually it flung them out into space, into the colorless, tasteless, weightless sea of outwardness without end.
>
> It flung them like stones.
>
> These unhappy agents found what had already been found in abundance on Earth—a nightmare of meaninglessness without end. The bounties of space, of infinite outwardness, were three: empty heroics, low comedy, and pointless death.
>
> Outwardness lost, at last, its imagined attractions.
>
> Only inwardness remained to be explored.
>
> Only the human soul remained *terra incognita.*
>
> This was the beginning of goodness and wisdom.

Those who agree with Vonnegut may not appreciate this trip. No deep law dictates that we must all share the same dreams or move in the same direction. For those who feel that there may be meaning out there, I would like to suggest two different purposes that we can follow.

Dreams for the New Millennium

1. We shall search for life forms, different from our own, within our solar system.

Philosopher Paul Davies has written that "there is little doubt that even the discovery of a single extraterrestrial microbe, if it could be shown to have evolved independently of life on Earth, would drastically alter our worldview and change our society as profoundly as the Copernican and Darwinian revolutions. It could truly be described as the greatest scientific discovery of all time."

Why should this be so? Such a discovery would settle one of the deepest questions that confronts us: whether the universe is innately fertile or barren. Two groups with very different worldviews agree on a key point: that the universe, by its nature, is a vast wasteland. One group attributes our existence in this desert to a divine act of creation, the other to blind luck operating against incredible odds.

The discovery of a second separate origin of life within a single solar system would strongly suggest that the laws of nature include some principle (I shall call it the Life Principle) that favors the generation of life. The same laws, in other contexts, help drive biological evolution and the development of societies. The entire history of the universe can then be interpreted in terms of a process called Cosmic Evolution. A new vision can be built around this concept that provides a sense of purpose for our own future.

The weakest point in this belief structure is our lack of understanding of the origin of life. No evidence remains that we know of to explain the steps that started life here, billions of years ago. If a broadly based life principle exists, however, then some signs of its handiwork should be detectable in the other worlds of our system.

With the exception of the Viking mission to Mars, this goal has been given short shrift in planetary exploration and until recently has been written off by most scientists. Certainly, the regions of the Moon that the Apollo astronauts explored were utterly dead—devoid of air, water, and organic substances. They can serve as a benchmark whenever a sample of truly sterile soil is needed.

But the remainder of the solar system is another matter. Admittedly, there is not much evidence for intelligent life there, and for most of the people with whom I have talked on this topic, alien intelligence is the most important question in outer space. But the detection of even the simplest life forms unrelated to us would define the universe as one in which we are, in biologist Stuart Kauffman's words, "at home" and "expected." Furthermore, this would be a universe that we are likely to share with intelligent companions.

2. We shall expand our biosphere beyond the Earth through human colonization of our own solar system and eventually of others.

If we value human life and recognize it as the product of billions of years of evolution, then we will want to safeguard it from extinction and secure it for the indefinite future.

The history of life on this planet has been punctuated by a series of catastrophes that have destroyed many of the species on it. A lesser disaster might spare us physically but destroy the basis of our technological civilization and our ability to proceed in that direction again. Even if we avoid the crises that have been abundant in the past, we will face one of much greater magnitude when our Sun runs out of fuel a few billion years from now.

As long as our species, and the others on which we depend, is confined to the surface of a single planet, we remain highly vulnerable. We will gain some insurance as soon as we establish a self-sustaining base somewhere off the Earth. We will be much safer yet when some of us have established footholds in other solar systems. This expansion off our crowded planet will also afford us many opportunities to explore what human existence can be and what we may choose to become.

Through Apollo, we have seen explorers in circumstances that were unlike any in our previous experience. These explorers were loping about under low gravity, with a blazing sun and black sky overhead, with the sites of all of our past history reduced to the dimensions of a ball in the sky. To many of us those moments symbolized the human future. They deserved the poems, the headlines, and the plaque on the library wall.

All of this deserves a greater exposition than a brief summary allows. To begin, we shall examine how this new dream contrasts with others that have been held worthy.

2

A Shift in the Cosmos

Stories

> Jack and Jill went up the hill
> To fetch a pail of water.
> Jack fell down and broke his crown
> And Jill came tumbling after.
> —*Nursery Rhyme*

We have each shared one common human experience: the slow emergence of our consciousness. As our senses took in a trickle and then a flood of information, we needed to arrange the information in some way that had meaning, that would guide us on what to do in this overwhelming world that surrounded us. As children, we would turn to our parents or relatives and bombard them with questions. A caring responder would often reply with a story, usually one framed to reflect the values and traditions of the community. The importance of stories has been emphasized by a host of psychologists, such as Jerome Bruner and those I have listed in parentheses after the following quotes:

> Stories are habitations. We live in and through stories. They conjure
> worlds. We do not know the world other than as story world. Stories

hold us together and keep us apart. We inhabit the great stories of our culture. (M. Mair)

Our interest in stories does not end with childhood.

Millions of people voluntarily spend countless hours listening to, reading or watching narratives. Most of television consists of stories of one sort or another—from soap operas to the Super Bowl. The popularity of novels, romances, comic books, mystery stories, and westerns again demonstrates how widespread is the human need for narrative. Anthropologists have long observed the popularity of stories in every culture. (Paul C. Vitz)

Many of us find personal meaning by creating our own stories to tie together the important events in our lives:

Our plannings, our remembering, even our loving and hating, are guided by narrative plots. (T. R. Sarbin)

Cultures behave similarly. Virtually all of them have had their own creation myth—an account of how the universe, living things, and the community came into being. This tradition continues today:

For example, a child's simple question of "How did the world start?" could be answered either by summarizing Genesis or by offering a thumbnail sketch of the Big Bang theory. (George S. Howard)

In Chapter 1, I argued that the Apollo mission had contradicted parts of the visions of past cultures but had failed to provide a perspective of its own. The conflicting versions of the lunar landscape that we examined, of course, were part of a much bigger picture: how those cultures saw the universe and our place in it. No one claims any longer (I think) that the Moon is inaccessible to us or is populated by bat-people, but other parts of those historic myths exist and affect our attitudes today. To penetrate the psychological and biological issues that are still at stake in planetary exploration, we will want to compare two worldviews. Most relevant are the one that science presents of the universe today and the one that preceded it, in medieval Western culture.

The Universe: A.D. *1300*

I have selected this time because a detailed, well-elaborated view of the cosmos existed. I would have preferred to use the year 1000, as it makes an apt contrast with the current millennial milestone. At that earlier date, however, the cultural image of the universe was somewhat like that of the Flat Earth Society. The Earth was platter shaped, with Jerusalem at the center and the arch of Heaven above. Conflicting observations, whether ancient or contemporary, went onto the trash heap. By 1300, links of commerce and conflict with the Muslim world had reintroduced the almost forgotten works of antiquity into the Western world. In particular, the astronomy of Aristotle and Ptolemy was found, in large part, to be consistent with biblical teachings, and a synthesis of the two was possible. St. Thomas Aquinas (1225–1274) was particularly effective in defending the reintroduction of Aristotle. A detailed, and in some aspects scientifically accurate, picture of the universe resulted. The picture that I describe in the following paragraphs has been taken from a number of sources listed in the notes, as well as the *Divine Comedy* of the great Italian poet Dante Alighieri (1265–1321).

In the late medieval worldview of A.D. 1300, the Earth was at the center of a limited and understandable universe. A number of ancient Athenians had reasoned that the Earth must be a sphere, and now it was seen as a sphere again. Its estimated size was not far from its current dimensions. Ptolemy had estimated that the Moon was 60,000 to 120,000 kilometers away, and this number, reasonably close to that of today, came into fashion again. Farther away yet were the planets, with the farthest boundary of the sphere of Saturn placed at a distance of 118 to 200 million kilometers. We now estimate the distance from the Sun to Saturn's orbit as about 1.42 billion kilometers. The order of the planets, proceeding out from the center of the solar system, was Earth, Moon, Mercury, Venus, Sun, Mars, Jupiter, and Saturn, which was correct, except that the Earth and its Moon had traded places with the Sun. So far, so good. But here, past and present versions diverge.

The preceding information came from Hellenistic science, but the ultimate base for knowledge in the fourteenth century was the Holy Scriptures. The remainder of the universe was described to conform to the requirements of the Scriptures. Places were set aside for the residence of the angels and of God, as well as for Purgatory and Hell. But let us return to the planets for a moment.

In the system of Ptolemy, the planets were not worlds, but symbols embedded in revolving transparent shells above the Earth. And so they remained in the late medieval cosmos, each representing an angelic being who lived in Paradise. The planets were simply denser regions in the spheres, and both were made of aether, a more noble building material than the substances used to construct our own world. The shells were in contact with one another but moved at varying speeds, without friction.

An additional eighth shell beyond Saturn contained the fixed stars. The total number of shells to this point was much less than the number used by Ptolemy, but the medieval era was more interested in clarity of concept than in accuracy of prediction. Philosophers disagreed on the number of shells that lay beyond the one that held the stars, as this question depended on interpretation rather than observation.

In one common version, the ninth shell was reserved for the Prime Mover, the force that kept the planets in motion. The tenth shell, sometimes called the Empyrean, was the home of the angels and of God. Here the universe ended, with not even empty space beyond. In the words of philosopher Arthur O. Lovejoy, "But though the medieval world was thus immense, relative to man and his planet, it was nevertheless definitely limited and fenced about. It was therefore essentially picturable; the perspectives which it presented, however great, were not wholly baffling to the imagination. The men of the fifteenth century still lived in a walled universe as well as in walled towns."

If we descended below the sphere that housed the Moon, we would find others made of air, fire, and water surrounding the Earth. In Dante's view, the Northern Hemisphere was inhabited, but the Southern one held no living humans; the only land in that hemisphere was the Mountain of Purgatory.

Our planet was made of baser materials and contained more corrupt beings than did the heavens. It was, however, the site of the most important action, where the forces of Heaven and Hell contended for the souls of individual human beings. The action of each person, in attaining salvation or causing damnation, sealed his or her fate for all eternity.

Those who were damned went into Hell, which existed below the surface of the Earth. As described by Dante, Hell was divided into nine different zones, each reserved for a particular type of sinner. In the First Circle, for example, we would find those who lived lives of virtue, but without faith in Christ. For example, Socrates and Plato, who preceded Christ, resided

there. As we descended through the shells toward the center, we would observe that the offense needed to earn that location became more severe,
and the punishment worse. Those who committed carnal sins inhabited
the Second Circle, but thieves, magicians, and false flatterers were much
farther down, in the Eighth Circle. In the Ninth Circle of Hell, the souls of
those deemed the worst of sinners, such as Judas, Brutus, and Cassius, were
tormented. The throne of Lucifer himself also sat in that Circle, which was
termed the Center of the Universe. We thus had a "diabolocentric" universe, one with the Devil at the very center.

The cosmos was limited in time as well as in space in the late medieval
view. Its creation by God over a period of six days is described in the Book
of Genesis. Light and Darkness were created above the waters on the first
day. On the second, Heaven was made. The dry land of the Earth was separated from the waters on the third day, and the land produced grass,
herbs, and fruit trees. On the fourth day, the Sun, Moon, and stars and
planets were created, and on the fifth winged creatures and those that live
in the seas. Creation was completed on the sixth day, when cattle, creeping
things, and the other beasts that live on land were made. Finally, man was
created and given dominion over all other living creatures. On the seventh
day, God rested.

The Bible was also used as a source book for the remainder of history.
That book does not provide exact dates, but it does list generations and
ages at which individuals died. This information was used to make a rough
estimate of the time when Creation occurred. It was assumed in the Middle Ages that the Earth was only a few thousand years old. In the seventeenth century, Archbishop James Ussher and other scholars consulted a
variety of Hebrew sources and concluded that Creation had taken place
about 4000 B.C.

The Earth's future duration was not expected to exceed its past history.
The Second Coming of Christ was anticipated in the near future. At that
time, the Devil would be imprisoned for a period of 1,000 years while
Christ ruled on Earth. Satan would then be freed and with his followers engage the forces of God in a final battle, which Satan would lose. The Earth
as we know it would then end, and the dead would all be resurrected for
the Last Judgment.

No date was set for the Second Coming of Christ, but in Matthew
16:28, Jesus is quoted: "Verily, I say unto you. There be some standing
here, which shall not taste of death, till they see the Son of man coming in

His kingdom." Many subsequent prophecies have been made of the Second Coming. One early Christian tradition, as described by Stephen Jay Gould, believed that each thousand years of human history was linked to a day of Creation. Six thousand years after the Creation would be the time of Christ's return. His reign on Earth would then represent the seventh day of Creation. If we accept Archbishop Ussher's exact reckoning, then that day came up on October 23, 1997, one which passed without any event of that magnitude. Some error in calculation was possible, of course. John Lightfoot of Cambridge University, a contemporary of Ussher, calculated that Creation occurred in September 3928 B.C. If Lightfoot was correct, then the Second Coming would be expected late in the twenty-first century.

Although the date of the Second Coming could remain uncertain, the major teachings of the Bible and the Church about the cosmos were accepted without dissent by the faithful in medieval times. As described by William Manchester in *A World Lit Only by Fire*, "There was no room in the medieval mind for doubt; the possibility of skepticism simply did not exist. *Katholikos*, Greek for 'universal,' had been used by theologians since the second century to distinguish Christianity from other religions."

The spirit of the time has been captured by Umberto Eco's novel entitled *The Name of the Rose*. This work is set in the year 1327 in an Italian abbey in which every conversation, every quarrel, debate, and conspiracy are placed within the accepted common ground of total faith in Holy Scripture. For example, the narrator, a novice visiting the abbey, tries to describe his infatuation with a young girl:

> Just as the whole universe is surely like a book written by the finger of God, in which everything speaks to us of the immense goodness of its Creator, in which every creature is description and mirror of life and death, in which the humblest rose becomes a gloss of our terrestrial progress—everything, in other words, spoke to me of the face I had hardly glimpsed in the aromatic shadows of the kitchen.

This belief system dovetailed well with an era when life was incredibly squalid for most of the population. Daily life in medieval times has been described by many historians. The account that I remember most vividly comes from a work of science fiction by Connie Willis, in which the circumstances of those times are described through the eyes of Kivrin, a time traveler from the twenty-first century. Kivrin has partly recovered from pneumonia and leaves the manor house where she is a guest to explore a nearby village:

> The track was muddy and rutted. . . . The cow moved off the green, its head down, into the shelter of the huts. Which were no shelter at all. They seemed hardly taller than Kivrin and as if they had been bundled together out of sticks and propped in place, and they didn't stop the wind at all.

Kivrin approaches a hut:

> It looked more like a haystack—grass and pieces of thatch wadded into the spaces between the poles, but its door was a mat of sticks tied together with blackish rope, the kind of door you could blow down with one breath. . . . It was dark inside and so smoky Kivrin couldn't see anything. It smelled terrible, like a stable. Worse than a stable. Mingled with the barnyard smells were smoke and mildew and the nasty odor of rats. . . . There was nowhere to sit in the hut . . . but the table had a wooden bowl and a heel of bread on it, and in the center of the hut, in the only cleared space, a little fire was burning in a shallow, dug-out hole. It was apparently the source of all the smoke even though there was a hole in the ceiling above it for a draft.

This description was meant to reflect daily life in an English village in the 1340s. Things could get worse yet, and they did. In the culmination of the novel, the Black Plague descends on the village and wipes out the entire population.

For those who had to suffer an existence of crushing poverty and hardship, their faith was a source of comfort. It promised that this life, however unpleasant it might be, was only a temporary period of trial before (it was hoped) the eternal pleasures of Paradise. Even the place of trial, the Earth, was only a temporary structure set up like an enormous tent, to be taken down after about 200 generations of humans had passed through it. Such was the view when our millennium was one-third complete.

Even then, however, another tradition was reawakening, that of science. The renewal of contacts with the East had brought back not only the astronomy of the ancient world, but all manner of practical learning. This is illustrated in another conversation in *The Name of the Rose*. The novice's master, Brother William, is challenged on his use of simple reading glasses. "What a wonder. And yet many would speak of witchcraft and diabolical machination." William defends their use:

> You can certainly speak of magic in this device. But there are two forms of magic. There is a magic that is the work of the Devil and which aims at man's downfall through artifices of which it is not licit to speak. But there is magic that is divine, where God's knowledge is made manifest through the knowledge of man, and serves to trans-form nature. . . . And this is holy magic, to which the learned must devote themselves more and more, not only to discover new things but to rediscover many secrets of nature that divine wisdom had revealed to the Hebrews, the Greeks, to other ancient peoples and even today, to the infidels.

Rediscovered science was first set firmly within the context of Holy Scripture, but as the millennium progressed, science increasingly took on an identity of its own. In the past several centuries it has totally reshaped the living conditions of most of the inhabitants of this planet, and it has greatly altered our view of the physical universe around us. We will want to compare this new cosmos to the late medieval one. But first, it is worth-while to say a few words about the rules of science. I will be brief here; a longer description can be found in my earlier book, *Origins,* and in two references I have cited in the Notes.

Science: The Method Is the Message

The question of rules comes up first because method is the essence of science. In the case of a myth, the story is the point. In Dante's *Divine Comedy,* Beatrice explained that the spots on the Moon represented different virtues in Heaven rather than a difference in height or a flaw of some sort. This explanation fit in with the larger theology of the times, in which the heavenly bodies were presumed to be perfect. The poetic statement was set forth as a final declaration rather than to provoke further dispute.

When Galileo constructed his telescope and turned it toward the Moon, he did so in a different spirit. Rather than deliver instructions to nature as to how the Moon must be, he was willing to let nature instruct him.

Had Dante been correct, Galileo would have seen a flat, lustrous surface in two hues. But what the telescope revealed was an irregular surface with mountains and craters. Galileo did not suppress his observations because they disagreed with previous dogma (although, as we shall see, the Church attempted to do this for some of his conclusions). Nor did his conclusions become dogma on their own. In the method of science, as formulated by philosopher Karl Popper and others, statements must be negatable, capable of being proven wrong.

Galileo, for example, also decided that the dark areas of the Moon must be seas. He got this part wrong. Lunar observations taken after Galileo, and confirmed in person by the astronauts, demonstrated that mountains and craters existed on the Moon, but seas did not (otherwise, Tranquility base would have been underwater). Galileo's observations on the mountains have survived, but his claims concerning the seas have been discarded.

In science, when two theories disagree, there must be a way to settle the dispute decisively. Ideally, a good scientist should be able to specify an experiment that would prove his or her theory wrong. He or she would be willing to lose if the results worked out the other way. One idea is withdrawn and the other rules the field, until the next contender appears. In this way science differs from religion, art criticism, ethics, and certain other areas, in which parties that disagree may broadcast their views indefinitely, without notable progress.

Observations, of course, can also be wrong. In Chapter 9 we will see how telescopic viewing errors convinced the astronomer Percival Lowell that artificial canals existed on Mars. An applicable rule of common sense is that extraordinary conclusions deserve extraordinary evidence. Observations that, if true, would change many paradigms of science or alter our entire view of the universe require massive confirmation. If I told my neighbor that I had seen a squirrel in my back yard, I would expect him to believe me just on the basis of my word. But suppose I had the notion that I had watched an alien spaceship land behind my house, and gray extraterrestrials had come out and invited me to tour their ship. Before I announced my experience to the world, I would want to be sure that I had borrowed one of the aliens' instruments or starmaps as proof of my story.

Science is progressive. As we pay greater attention to what nature has to tell us, we gain greater ability to predict events and to control our ultimate fate. But science gives us no instructions concerning the use of our new power or our ultimate purposes. If we want a guide for our future direction, then we need a new story, one that agrees with our new picture of the universe. Before we tackle that task, we need to examine the stage on which science has placed us.

The Museum of the Cosmos

The vision of the universe produced by science at the end of the millennium is constructed on a scale so large that we run into a communication problem in attempting to describe it. For example, the closest star, Alpha Centauri, is about 40,000,000,000,000 kilometers away. Even the most frequent flyer among us would find that that number reduces his lifetime mileage to insignificance. And if the next star over is so incredibly far away, how do we compare its distance to the much larger distances that separate galaxies? To make such numbers meaningful to us, we must convert them to a form that we can understand in terms of our usual daily experience.

One clue as to how to begin comes from a shorthand used by scientists. Our eyes are not skilled in quickly adding up the number of zeroes in a long string of them, so we write such numbers as *exponents*. In this system the number 40,000,000,000,000 can be written, using much less space, as 4×10^{13}, where the raised number, 13, is called the exponent. In that term, 10^{13} represents the number 1, followed by 13 zeroes. The term can be read as an instruction: Write down 4 and follow it with 13 zeroes (putting in commas if you prefer it that way). The exponent, 13 in this case, may be considered as a zero counter.

Apart from saving space, the exponent system makes comparisons much easier. Suppose that I had asked the aliens in their spaceship for a ride to Alpha Centauri. They set out, but after a while they told me that their plans had changed, and they left me out in space (in a spacesuit, hopefully) at a bus stop. A roadsign that pointed back the way I had come said "Sol, 4×10^{12} kilometers." (Sol is a name used for our own Sun.) How much of the distance would I have covered? I would quickly appreciate that the distance I had come had one less zero than the whole distance. The situation resembled one in which I had been given a ride of 4 kilometers toward a town 40 kilometers away. I had come only one-tenth of the way.

A shortcut in trying to understand distances that vary enormously in size is to focus on the exponent and worry less about the front part. For example, I have flown from New York to San Francisco a number of times, a distance of about 4,000 kilometers. This number could be written as 4×10^3 kilometers. I sometimes run in marathon road races, which cover about 40 (4×10^1) kilometers. We can compare the three distances: a marathon race, New York to San Francisco, and Sol to Alpha Centauri, by remembering the exponents 1, 3, and 13. If I ran 100 marathons (I won't, but I have a friend who hopes to do that), then I would have covered the distance to San Francisco. A number with the exponent 3 is 100 times (10^2) larger than the same number with the exponent 1. One hundred flights to San Francisco, however, would bring me only to exponent level 5. I would need 10 billion such flights (1 followed by 10 zeroes) to get to the nearest star.

We can use exponents to classify any distance in the universe, and we will find it much easier to keep things in proportion if we do so. But numbers are still abstractions. We experience things more fully when we can see them. I have two books on my bookshelf, *Cosmic View* and *Powers of Ten*, that use this idea to produce views of objects of different sizes, from the tiny nucleus of an atom to the universe itself. Each page is used to represent a different exponent, or level of magnification. In *Cosmic View*, for example, one page shows a man asleep on a lawn, another the whole Earth, while yet another contains most of the solar system.

It is very helpful to go from page to page and see things get larger in leaps of ten. But the feeling I get as I do this is roughly like the one I receive when I look at a travel magazine. A photo of a street in an exotic city holds my interest, but this falls short of the experience that I get when I am there in person. Somewhere in between lies the experience of wandering through a cleverly constructed model.

Imagine, then, a group of models that represent aspects of the universe at different levels of magnification, each ten times larger than the previous one. Rooms might do for some levels, while an entire floor of a large building would be needed to do justice to others. I fancy the idea that the models could be housed together within one large building, which we would call the Museum of the Cosmos. I hope that at least one is built someday, as a monument to our age of science.

The late Middle Ages left splendid monuments to their faith in the form of the great cathedrals. The cathedral in Chartres, built in the thirteenth century, towers about 107 meters over the flat terrain of north central France. It quite dwarfed the ordinary dwellings and places of business of the townspeople of the time when it was constructed. In our age, technology has allowed the construction of skyscrapers that far exceed the height of Chartres Cathedral. They are dedicated to business, however.

Science museums can be found in a number of cities, but they are not usually as imposing in appearance as the skyscrapers. Further, they represent the various areas of scientific learning separately and do not make a unified presentation of our current view of the cosmos. A Museum of the Cosmos could make a statement for our age and perform an additional valuable service: It would present the modern scientific view of the universe in a way that most human beings could understand. If this Museum were successful, it might be duplicated in a number of places, just as cathedrals were built in many towns in Europe. No such building exists as yet, so we must create one in our imaginations to help us absorb the vastness that lies out there.

A View of the City and of the Planet

We can start with a familiar scene: scale models of towns and cities. I can remember one such model of New York City displayed at the New York World's Fair in 1964. In our Museum, if we wanted to include all of New York City but still keep the display at a reasonable size, we would scale it at 1:1,000. Our apparent size would be 1,000 times (or 10^3) larger than our real one, or, if you prefer, the buildings would be 1,000 times smaller. We would call this location Level 3 to match the exponent.

I am about 1.8 meters tall, but on Level 3, my apparent height would be 1.8 kilometers. The tallest buildings would be shin high, and it would take me about fifteen paces to walk the length of Manhattan Island. It would be fun to explore such a model, but it wouldn't teach us much about the universe. We must move on.

Some New Yorkers feel that the world ends just outside the city limits. Here we are chasing scientific, not poetic, reality, so we will leap up to Level 7 model for a convenient view of our globe. In this model, my apparent height would increase by another 10,000 (10^4) times, so that I would be a bit larger than our planet, while Manhattan would now be about the size of the letter "i" on this page. For dramatic effect, I would want the Earth to hang suspended before me in all its beauty, looking just like the photos taken from space. The Moon would be represented by another ball, about one-quarter the size of Earth in its diameter, but for accuracy, we would have to place it about half a city block away. No other worlds could be conveniently included in this room because the nearest planet, at its closest approach, would still be over 4 kilometers distant.

Lots of Space: Our Solar System and Its Neighbors

In the medieval cosmos, Saturn was the outermost planet. If we wanted a model for our Museum that displayed the orbits of the planets out to Saturn in a reasonable space, we would have to use the dimensions of Level 11, where objects are reduced to one hundred billionth of their size (or we are scaled up by the same amount). At that level, the orbits of all of the planets known to the ancients could be fit into a display area about the size of a baseball diamond. But there would be very little to see.

The largest object would be the Sun, which would be represented by a bright light source about the width of a fingernail. The planets would be much less obvious, and we would need illuminated arrows to guide visitors to their locations. Jupiter, the largest of the planets, is about one-tenth the diameter of the Sun, so it would be a tiny sphere about the size of the letter "o" on this page, and Saturn would be slightly smaller. We would find the Earth about an arm's length from the Sun, but it would be only the size of a dot and barely visible with the naked eye. The major impression that we would get at this level would be the emptiness of space, even within the realm of our solar system. Things would get worse as we climbed up to higher levels.

Suppose we had the urge to build an exhibit that included our Sun and some celestial neighbors. As we mentioned earlier, the Sun to Alpha Centauri distance could be represented in kilometers by the number 4 followed by 13 zeroes, or 4×10^{13}. In our Museum, we express distances in meters, so that number will be larger after we convert kilometers to meters. The Sun to Alpha Centauri distance in meters becomes 4×10^{16}. Astronomers prefer to express such huge distances more simply. In one year, light will travel just under 1×10^{16} meters in a vacuum. That distance is described as 1 light-year. The light from Alpha Centauri takes about 4.3 years to reach us, so we say that the star is 4.3 light-years away.

At Level 16 of the Museum, a light-year is represented by a distance of about 1 meter. If we made our room fairly large, we could fit in a few dozen stars in addition to the Sun and Alpha Centauri. Among them would be

Sirius, the brightest star in the sky, which is 8.8 light-years away. Unfortunately, all of them would be too small to be seen, except as points of light. Alpha Centauri is actually a triple star, with two brighter companions and a dim one in orbit around one another, and we could illustrate this by placing separate dots of light a few centimeters from one another. Perhaps a few artificial markers could be included to relieve the starkness of a room containing only a scattering of tiny bright pinpoints. For example, if we drew the orbit of the outermost planet of our own Sun, Pluto, that path would be just visible as a tiny circle.

The major point to be absorbed at this level, however, is that we have broken entirely out of the plan of the medieval cosmos. The stars are not embedded in a dome above us (except when we visit a planetarium) but are widely scattered in space. Their separation is so great that a walkable model that includes a few of them shrinks our solar system almost to invisibility. But at this level, we have not even started to come to terms with the dimensions of the universe. We will jump up four levels for another look.

The Galaxy Room

If a Museum of the Cosmos is ever constructed, Level 20 will be one of its most popular attractions. Imagine a room whose center is occupied by a magnificent disk-shaped object, 10 meters in diameter, so large that in reality light takes about 100,000 years to cross it. This model represents our home galaxy, which we call the Milky Way.

Our model would contain a center of glowing light surrounded by several illuminated arms that spiral out from it, providing a pinwheel effect. The objects that produce the light are individual suns, 100 billion of them, but they would be much too small to be seen separately. At this level, the contents of the room that contained our Sun and Alpha Centauri at Level 16 would now be compressed within the size of the letter "o." We might choose, however, to put up a marker to show the location of our own solar system. It would be embedded in a spiral arm, some 30,000 light-years out from the center. And the entire object would be framed by the utter blackness of space. The glowing clouds that comprise the arms are not uniformly bright, however. They are marred, here and there, by marble-sized darker patches, which represent regions of gas and dust of the type that give birth to our Sun and to other stars.

When we look at the actual sky at night, we find that our location in a spiral arm gives us a good view toward the galactic center. We see that center as the diffuse glowing cloud in the sky that we call the Milky Way, a name later applied to the entire galaxy. Galileo turned his telescope in that direction in 1609 and observed "innumerable stars grouped together in clusters." Dust clouds obscure our view of the very galactic center, but many scientists believe that a large black hole (an object with gravity so strong that nothing, not even light, can escape from it) may lie there in a quiet state.

If the budget available for the construction of our Galaxy Room was generous enough, we could include additional objects. Smaller satellite galaxies, the Large and Small Magellanic Clouds, lie 170 and 200 thousand light-years from the center of the Milky Way, or at about twice its diameter. A model of another full-sized galaxy, one that is close to being a twin of our own, would be placed at ten times the distance of the Small Magellanic Cloud, or 2 million light-years away. We would have to enlarge our Galaxy Room to 200 meters to include it. This sister galaxy, which has been known as the Andromeda Nebula, appears as a fuzzy patch of light in the nighttime sky. It is the most distant object that can be seen with the naked eye.

At the start of this century, astronomers felt that our galaxy constituted the entire universe. Only in the 1920s did Edwin Hubble and others establish that the Andromeda Nebula was a separate one. It now bears the more formal name of Messier 31 or M31. The Milky Way and M31 are both part of a local group of 26 galaxies that contains one other major member, M33 by name, and smaller ones. The galaxies, as we have seen from our visit to Level 20, are much closer together relative to their size than are planets and stars. As if in compensation, however, the empty space between the galaxies is far, far emptier of isolated atoms than is the space between the stars.

We have seen enough to overwhelm us, but we have covered only a tiny segment of the universe. We have to visit one more level, one in which we can view (with apologies to Timothy Ferris) *The Whole Shebang*.

The Universe Room

When we consider how to model Level 24, we face conceptual problems that did not come up at the lower ones. We will want to reduce the universe to a scale where 1 meter equals the distance that light travels in 100 million years. Our model will display the closer areas that have been mapped in some detail, as well as some isolated features that are much farther away. But a strict rule will limit the distance at which we place the farthest object in the room: Its distance in light-years cannot be greater than the time, in years, since the universe was created. If the object lay outside that realm, then its light would not have reached us and we could not observe it. Our model must be limited to the observable universe.

How large a room will we need? Unfortunately, astronomers do not agree on the age of the universe; estimates currently range from 8 to 18 billion years. Not too long ago, Professor Sallie Baliunas of the Harvard-Smithsonian Center for Astrophysics showed me a chart in which the various estimates were gradually converging at 12 billion years. If we adopt this number, we can specify that the room will extend no more than 120 meters from the center in any direction. We would still require an arena the size of a sports stadium, but a grand subject deserves a grand display.

A different problem now emerges: What object should we place at the center? By definition, the observable universe must center on the observer, so we will place the center right here on Earth. Copernicus may have removed our planet from the center of things, but the Universe Room has put us right back there again. On our Level 24 model, of course, the entire Milky Way Galaxy, with its hundred billion stars, could claim that honor, as our entire galaxy would be represented by a glowing object a bit smaller than the letter "o."

At present, we cannot specify the contents of the room in any detail. Only limited portions of the cosmos have been mapped, and available telescopes do not reach as far as the theoretical barrier at the speed of light. Certain limited structures could be displayed, and the remainder of the area would be left in dim light or not defined at all.

In the region that has been explored we would see hordes of tiny bright dots representing galaxies. In one study in the mid-1990s, the Hubble telescope was pointed at one very narrow segment of sky near the handle of the Big Dipper. In that small sample, 1,500 to 2,000 galaxies were counted. By extrapolation, it was decided that the total number in all directions might include as many as 50 billion. If we stared closely at our exhibit, we could recognize that galaxies occur in several shapes: spirals (like our own), ellipses, and irregular forms. These bright smudges would not be scattered uniformly through space, however; they would be arranged into definite structures. The smallest unit of structure contains a moderate number of galaxies arranged in a group, such as the one that includes the Milky Way. A group will usually have three to six prominent members and a dozen or so smaller hangers-on. On Level 24, a collection of this sort might occupy a space as large as your fist. The groups, in turn, are organized into clusters of hundreds to thousands of galaxies. Our local group is part of a larger unit, named the Local or Coma-Sculptor Cloud.

One prominent landmark some 60 million light-years away has been named the Virgo cluster, after the constellation in which it occurs. According to Timothy Ferris, the heavenly pathway leading from our galaxy to the Virgo cluster is a "Louvre gallery of picturesque galaxies." If we choose to travel farther on the same path, we will enter an area that holds a greater mystery. A large and massive object is pulling both our local group and the Virgo cluster in its direction. We cannot see what is tugging us, because dark clouds of the Milky Way partly obscure it and also because most of it is made of "dark matter," a type of substance not yet identified. This object, 200 million light-years away, has been named the "Great Attractor" by Alan Dressler and other astronomers.

Both groups and clusters are held together by gravity, but larger units are not; they reflect the earlier processes that shaped the universe. The clusters congregate in superclusters, which in turn form enormous lacy sheets of galaxies, with names such as the Great Wall and the Cetus Wall. These sheets can extend for as much as 1 billion light-years, or about 10 meters on Level 24. They enclose voids or bubbles where galaxies are rare, giving a spongelike texture to the cosmos.

We will need some objects to decorate the farther reaches of our model, and quasars are the most likely candidates. These extremely bright objects can emit more light than a trillion suns, and in about 10 percent of the cases they are potent radio emitters as well. Most astronomers believe that the radiation is released by gases and entire stars that are in the process of being sucked into an enormous and active black hole. The majority of these objects appear at distances many billions of light-years away, so they can be represented as bright points of light in the more distant parts of our Universe Room. The quasars were formed through processes that were much more active in the early universe than now, so we see them mostly in the distance because of another effect that operates in our model and the reality that it represents.

A Long Time Ago

The film *Star Wars* starts with the phrase, "A long time ago in a galaxy far, far away." In practice, whenever we observe a galaxy that is far away, we must see it as it was long ago. Light from the Virgo cluster, for example, started on its trip to us 60 million years ago. When it finally reaches us, we see the cluster as it appeared at the end of the reign of dinosaurs on Earth. As we stroll through Level 24, moving from the Milky Way Galaxy toward the edge of the exhibit, we are also traveling through time, observing objects as they were much earlier in their existence. We can see how some very distant galaxies appeared when the universe was less than 10 percent of its present age. They are quite ancient now, but we can only view them in their infancy.

Time and space are linked in our cosmos. We have completed our tour of space; now we must consider time. We will shift our focus and compare the scientific account of the history of the universe as presented by science with the earlier biblical version.

Creation Week (Revised Version)

The Book of Genesis describes a Creation period that lasted six days, with a seventh day devoted to rest. In the modern scientific view, this process took billions of years. But it is almost as hard to cope with billions of years of time as it is to comprehend distances of billions of light-years in space. In some religions, the "days" of Creation described in the Bible are considered to be metaphors for much longer periods of time. We will reverse this idea and describe a 12-billion-year history of the universe as though it were taking place over six days, starting just after midnight on Monday. Each day will correspond to 2 billion years. As our interest centers on the events that led to us, we will pass quickly over the first few days.

Most scientists believe that space, time, matter, and energy were created in an instant in the process called the Big Bang. Gas clouds of primordial matter then came together to form the first low-mass galaxies, which later coalesced to form larger ones. Our own Milky Way Galaxy probably started to come together within this period, and the first stars were also born. Monday was the busiest day of Creation Week. By Tuesday, stars were being formed in abundance within the galaxies. The new stars then entered into their life cycles, converting primordial hydrogen and helium to heavier elements. When the stars had completed their cycles and exhausted their fuel supplies, they dumped their contents into space, sometimes in a huge explosion called a supernova. The resulting clouds of dust served, in turn, as building material for new generations of stars, enriched in the heavier elements.

Those were exciting times in the universe. As Timothy Ferris described it, "the youthful universe was ablaze with starburst galaxies." After about 4 billion years, however, the rate of star formation began to drop dramatically, having peaked at about 12 times its current rate. Astronomers now feel, according to Ferris, "that cosmic history resembles a party at which we have arrived only after the fireworks show has ended, the choice food and drink [have] been consumed, and most of the guests have faded from view." In the words of James Peebles of Princeton, "Things are running down."

Much of the action had shut down by Wednesday evening of Creation Week, but one event that matters to us had not yet taken place: the birth of our solar system. The "observable" universe was smaller then, extending only 6 billion light-years in all directions, but we were not there to observe it. In the place of the Sun and Earth and the other familiar worlds, there was at best a pregnant gas cloud. It was only late on Thursday afternoon, some 4.6 billion years ago, that our solar system was formed.

We were not there to watch it, of course, but we can learn about the birth of our Sun through calculations and by observing similar events that are taking place elsewhere in our galaxy. The Orion Nebula, for example, is busily engaged in star formation, and an existing young star, MWC480, may be giving birth to planets at the present time.

Solar system formation begins when a cloud of gas and dust is disturbed by a nearby supernova or some other provocation and starts to fall together by gravitational attraction. A flattened, spinning disk with a central bulge is formed, called the solar nebula. Most of the material collects at the center and heats up due to the release of the energy of gravity. After a certain mass and temperature are reached, a nuclear reaction begins, converting hydrogen to helium. When this long-lasting energy source switches on, a star is born. Not all of the material collects in the center, however. Additional debris remains in orbit at various distances and gathers to form planets, asteroids, and comets.

Our Earth is likely to have formed in this way, over a period of 10 to

100 million years. Most of the material probably arrived in bits and pieces, but collisions with larger bodies probably contributed as well. As we mentioned earlier, an encounter with a planet larger than Mars probably led to the formation of our Moon. Further impacts by large bodies kept the Earth's surface in a molten state and helped the interior settle out. The heavier molten iron sank to the center to form a core, and the lighter rocky materials floated to the top.

We cannot be sure when the heavy bombardment stopped. The lunar crater record testifies to a final wave of cataclysmic impacts at about 3.9 billion years ago. This may have affected the Earth as well or only reflected local events on the Moon. Afterward, very early on Friday morning of Creation Week, something strange and special took place. Life began.

We have only a few rock formations to tell us of those times. They suggest an Earth with an atmosphere like our own today, but lacking in oxygen. Oceans existed at that time, with volcanoes protruding above them, and possibly some continental land was there as well, but with less of it above water than at present. The Earth may have been rotating more rapidly, affording shorter days. The Moon was closer to the Earth than it is today, producing enormous tides that swept the shorelines. The young Sun was about 30 percent less bright than at present, which would have led to a much colder planet if the following process had not intervened. Some scientists believe that the carbon dioxide levels in the air were higher at that time, creating a "greenhouse effect" in which heat was retained in our atmosphere more efficiently and the oceans were saved from freezing.

The evidence that life was present then is rather weak—hardly the "smoking gun" that the media talk of in murder trials and political scandals. No jawbone or footprint marks the ancient minerals. But carbon atoms remain. This element plays a crucial role in life today, but it also exists in many substances that may have nothing to do with life (for example, limestone rocks on Earth and bizarre molecules in outer space). We cannot conclude that life was present in an environment just because we detect simple carbon compounds there.

Carbon atoms have a quirk, however, by which they can reveal something of their past histories. They can occur in two stable forms (called isotopes), and the ratio of the two isotopes formed when carbon has been deposited by mineral processes differs from the ratio produced by living organisms of the type we know today. My friend Gustaf Arrhenius and his co-workers have collected and analyzed material from 3.85-billion-year-old rocks on bleak Akilia Island, off the southwestern coast of Greenland. The carbon isotope analysis suggests to them that life was present at that time. Some scientists have been concerned that the interval between the end of the heavy bombardment of the Earth and the formation of the Akilia Island rocks was too short for life to have started here. As we noted previously, however, the bombardment may have ended earlier. Further, we really have no idea of how much time was needed for life to begin.

For more abundant evidence of life we must turn to the deserts of northwestern Australia, where a series of microscopic fossil imprints has been found that date from 3.5 billion years ago. Nothing more than the shapes remain: chains of cells that in their overall dimensions resemble modern blue-green algae. Such organisms today belong to a class called prokaryotes: a formidable name that describes microbes with smaller, simpler cells and no nucleus. Most creatures apart from the blue-green algae and bacteria employ a much more complex type of cell; the name *eukaryotes* describes this group. We are members of this group, as are elephants, worms, daisies, yeast, and almost every other living thing that you can bring to mind.

Another type of evidence that testifies that life existed 3.5 billion years ago can be found not far from the Australian microfossil site. When blue-green algae and some other microbes grow in shallow waters, they form foot-high dome-shaped, highly layered structures called stromatolytes. These objects are produced when a film of the organisms gets covered with debris. A new layer of blue-green algae forms above the debris and gets covered in turn, until a multilayered cabbagelike structure develops. Stromatolytes exist today off the coast of Australia. Ancient fossil shapes, lacking the living algae but with the appropriate rock layers in place, testify to life at a much earlier time.

We need a word of caution at this point. Scientific accounts are meant to be doubted: this is part of the normal process. They differ in this way from myths, which are meant to be accepted on the word of the source. Stromatolyte-type structures can also sometimes be formed by processes not part of life. Minerals have a knack, unfortunately, of forming organized structures that at first glance may resemble a biological shape. The identification of microfossils is an area where doubt has been developed to a fine art. Let us quote evolutionist George Gaylord Simpson concerning a famous fossil of the nineteenth century: "Eozoon, proudly named the 'dawn animal,' is now considered to be no animal at all, nor yet a plant or any form of life but a mere inorganic precipitate."

I have brought up this particular point to give some sense of the care used in making the claim that life was present on Earth 3.5 billion years ago. We will want to keep these difficulties in mind when we consider the possibilities of locating microfossils on other worlds. In the case of our own planet, many claims of ancient microfossils have been rejected, but a few secure ones remain.

Life, then, was in business on Earth by dawn on Friday, the fifth of the six "days" of Creation Week. Without further knowledge, we might assume that events moved forward very rapidly after that. But existing information does not bear us out. Microfossils become more numerous as we move through Friday and into Saturday morning, less than 2 billion years ago, but they continue to resemble simple prokaryotes. Finally, early on Saturday, fossils resembling the more complex eukaryotes appear. One other noteworthy change in our planet took place, however, a bit before that time (about 2 billion years ago): Large quantities of oxygen appeared in the atmosphere.

Certain bacterial strains may have triggered this change in the air by inventing a new type of photosynthesis. They learned how to use light energy from the Sun to make sugars from carbon dioxide and water, and they released oxygen as a byproduct. In doing so, they triggered one of Earth's earlier environmental catastrophies, as this gas can be corrosive and toxic to organisms unequipped to handle it. A large-scale slaughter of oxygen-sensitive microbes probably resulted from this development. Some survivors still linger on to this day, however, in locations such as the mud at the bottom of the Black Sea and San Francisco Bay.

We humans, of course, are utterly dependent on the 20 percent of oxygen in the air. If you put a candle under an upside-down glass, you will see it snuff out when it has used up the oxygen inside the glass. A mouse kept within the glass would die for the same reasons, and human beings also die when they are trapped without oxygen. It is sobering to think that the world we call Mother Earth would have smothered us, rather than mothered us, if we traveled back somehow to any moment in the first half of its history. When we travel, many centuries from now, to Earth-like planets around other suns, we had best keep our spacesuits on, no matter how gentle the waves on the beaches may be.

We have reached noon on Saturday, the sixth day of Creation Week. Evolution on Earth has moved sluggishly to this point, though some more complex single cells and perhaps a few fossils of multicellular organisms did turn up on Saturday morning. Finally, late Saturday afternoon (about 530 million years ago), an enormous explosion of life took place in the seas. An abundant fossil record of shells and other hard parts document this time, which has been called the Cambrian era. Stephen Jay Gould has provided a vivid description of the strange array of backbone-less creatures that lived in the Cambrian seas. Most of them became extinct, but among the survivors was Pikaia, a minor species once mistaken for a worm. After a closer examination, however, it was reclassified. Pikaia had the makings of a primitive spinal cord and was an early member of the chordate group. Chordates later gave rise to the vertebrates, the class that contains backboned animals, including ourselves. Those of us who keep collections of photographs and paintings of our ancestors may wish to include Pikaia as well.

The final hours of Creation Week look much less strange to us, as fishlike creatures appeared, followed by plants and animals that colonized the land. The media have made the era of dinosaurs, which lasted from 220 to 60 million years ago, look almost as familiar as the contents of our local zoo. This period, during which birds and mammals first developed, lasted for only two hours on the scale of Creation Week. Finally, the evolution of humans from primate ancestors took place over the last five to ten minutes. The six-thousand-year period of modern human history that is described by the Bible and other histories occupies less than a second of the scientific version of Creation Week.

The history of life on Earth has been punctuated not only by the appearance of new species, but also by massive extinctions, which have occurred periodically. The most famous episode took place 65 million years ago and swept the dinosaurs away, together with about half of the other existing species. Many scientists believe that this massive slaughter was caused by an impact from a large meteorite or comet, one that created the 100-kilometer-wide Chicxulub crater near the Yucatan peninsula in Mexico. An even more lethal but less well known mass extinction struck some 250 million years ago, in the late Permian period. This disaster eliminated 90 percent of the species in the oceans and more than two-thirds of the existing reptile and amphibian families. Multiple volcanic eruptions, a sudden release of carbon dioxide from the depths of the oceans, and a comet or asteroid impact have been suggested as possible causes.

My tale of Creation Week has brought us to the present, after 12 billion years, which I have equated with six days. In the biblical account, the Creator used the seventh day for rest. In terms of our own future, it is hard to project the next century or millennium, let alone the next 2 billion years. But rest seems unlikely. For those of us who value the continued survival of the human race and its descendants, long-term activity is essential.

There is no reason to believe that the extinction cycles have ended. Comets should continue to arrive, and the other causes of catastrophes are likely to appear again. Our Sun may enter a cloud of interstellar dust in 50,000 years, with unpredictable effects. We can be more certain about the increasing luminosity of our Sun. In a billion years, the increasing heat will disturb geochemical cycles, and most of our water supply may be lost to space as a result.

An ultimate deadline for our activities in this solar system will come in 5 or 6 billion years (next Tuesday, if we are just now starting the Sunday of Creation Week). At that time our Sun will have run out of hydrogen fuel and will burn out. In the first stage of its death, it will turn into a red dwarf star, with clouds of hot gas extending far beyond its current borders. In the words of John Noble Wilford of the *New York Times*, "The expanding solar surface might swallow up Earth or stop just short of that. Even if Earth was not swallowed up, the Sun would fill half the sky. The heat would be so intense that it would scorch the planet's atmosphere, vaporizing vegetation and boiling away the oceans." Astronomer Howard Bond commented, "After the Sun has become a red giant and burned the Earth to a cinder, it will eject its own beautiful nebula and then fade away as a white dwarf."

The articles from which I took the preceding quotes were illustrated with Hubble telescope pictures of stars that are dying at the present time: The expanding gases made colorful patterns that resembled pinwheels, goblets, hourglasses, and cat's eyes. Which would our own Sun resemble? Whichever it was, I hoped that the humans of the time would watch it through another telescope. As Dr. Bond concluded, "We have another six billion years to get out of town." But the next town is very, very far down the road.

3

A Matter of Perspective

The scientific view of the universe that we outlined in the previous chapter stands in stark contrast to the medieval one. In the latter view, the universe was not tiny, of course. If we constructed a model of the medieval universe within a huge domed baseball stadium, the Earth would stand at the center, the size of a marble, with the shell of the fixed stars making up the outer walls of the arena. But when we used the same stadium to model the scientific universe in Chapter 2, the center was occupied by an object, our home galaxy, which was only one-tenth the size of that marble. Yet our Milky Way Galaxy far exceeds in size anything that our ancestors comprehended.

Imagine now that we swell up our galaxy to the size of the entire arena. An object the size of the medieval cosmos would not be visible to the naked eye or with a microscope. We would need an electron microscope to see it, as it would be about the size of a virus.

The scientific time scale also dwarfed the medieval one. We saw in the last chapter that the age of the universe stood in proportion to the period of recorded human history as six days do to about a quarter of a second.

To those who accepted the literal truth of the Holy Scriptures, these discoveries were unsettling. Physicist-philosopher Harold J. Morowitz has summed this up very well:

> To believers in the exact word of Genesis, man is the raison d'être of creation, and all the rest of the universe consists of a group of heavenly objects set there by the Creator to decorate the human abode. . . .

Judaism, Christianity, Islam and their offshoot sects are critically dependent on reports of certain past events in which the Lord or a representative agent of the Almighty communicated instructions to selected groups or individuals. . . . Either the events occurred as reported, or the bases upon which the respective theologies are founded simply disappear. If one begins to doubt part of the sacred texts, then the entire books somehow come into question. . . . Either their scriptures are absolute and unquestioned, or their claims to God's existence and his relation to them become suspect. Fundamentalists can be seen more clearly and perhaps more sympathetically in terms of this all-or-none choice, which has motivated and continues to motivate many believers.

When the new revelations of science were more limited in scope, the simplest strategy was to deny them. Clerical authorities protected not only the word of the Bible, but also the findings of Aristotle that were compatible with Holy Scriptures. The Church did not take kindly to the idea of Copernicus that the Earth orbits the Sun, nor did it approve of Galileo's lunar observations. Let me cite a summary of the charges against Galileo written by Tomasso Campanella in 1622:

> Galileo says that water exists on the moon and the planets, which cannot be. These bodies are incorruptible, for do not all Scholastics contend with Aristotle that they endure without change throughout all time? He describes lands and mountains in the moon and other celestial globes, and not only villifies immeasurably the homes of the angels, but lessens our hope regarding heaven.

The viewpoint of the Church has been interpreted for modern readers by playwright Barrie Stavis, in his work *Lamp at Midnight,* which deals with Galileo's ordeal after he first supported the system of Copernicus. In the following excerpt, Galileo's examiner, Cardinal Bellarmine, speaks for the Church:

> As Christianity developed, it became urgent to adopt a single official system of the universe. The Fathers of the Church found Aristotle most in accord with the spirit of scripture. For hundreds of years the astronomy of Aristotle and the heavens of Christian theology have

been as one. . . . I must ask only one question. What will happen to Christian teaching if our system of the heavens were to be torn down and your system set up in its place? And the answer is: Christian truth would be destroyed! . . . You would transform the Church of the entire universe into the church of one insignificant clod of dirt, lost in space.

Galileo was, in fact, requested not to teach the Copernican system as truth, and he obeyed this edict for some years. But after Cardinal Bellarmine died, he resumed his advocacy. In 1633, Galileo was forced to appear before the Inquisition in Rome. He renounced his beliefs and remained under house arrest until his death.

In subsequent centuries, the Church learned the folly of tying the validity of a religious system to ideas that can be negated simply by turning a telescope to the Moon. (Galileo, however, was not fully exonerated until the 1990s.) A better strategy in defense of the Christian belief system was simply to abandon Aristotle and accept the scientific view of the universe. The arrangement of the heavenly bodies in space is not central to the teachings of the Bible.

Relocation of some of the prominent parts of the medieval cosmos would be needed for this revision, however. In a recent poll by the Gallup and Barna Research Group cited by the *New York Times,* 73 percent of Americans professed to believe in the existence of Hell. Geologists now claim that the center of the Earth is composed of molten iron, which would be too hot even for Dante's conception of the Inferno. Although there is more than ample room in the modern universe to house both Hell and Heaven, my suspicion is that they have been moved to some other plane of existence entirely, far beyond the reach of any space probe.

Some sects, however, have continued to defend the time scale of the Bible while conceding to science on issues involving space. In 1997, I visited a museum maintained by Creationists to get a better idea of their interpretation of the cosmos.

Just off the Freeway

I had been visiting San Diego to attend the annual meeting of the American Association for Cancer Research. To reach the museum, I had to drive for an hour on freeways that led east from the coast to the town of Santee.

I traveled through arid rolling hills, which reminded me of the biblical countryside outside Jerusalem. I had hoped that my destination would lie atop a hillside surrounded by wilderness. Instead I found myself on the service road of the highway, which also contained industrial properties and retail outlets.

The building I sought was low and attractive, though, with ample parking. It housed the Institute for Creation Research and the Museum of Creation and Earth History. The latter claimed my immediate attention. It held a series of connected rooms, all on one floor, that were dedicated to a number of topics on the theme of science. An astronomy exhibit caught my attention almost immediately. Large NASA photographs from the Voyager mission displayed the planets in their full glory. Everything seemed very conventional until I noticed a placard that pointed out how different the planets were from one another. It concluded that no common process had produced them. Each was the work of a separate act of Creation.

A museum aide was guiding a group of school-age children through the area. When I entered the museum, I learned that these children had been kept out of public school by their parents, to protect them from materials that contradicted the Bible. They received home instruction, supplemented by excursions such as the current one. One of them now asked their instructor why stars sometimes exploded. The instructor's answer was that nobody knew for certain, and he supposed that they had suffered an internal disturbance of some sort before they blew up.

This answer disregarded a well-established area of astronomical knowledge. The characteristics of stars have been classified in a scheme called the Hertzsprung-Russell diagram, and the life histories of stars of various sizes have been worked out in detail. Stars much more massive than our Sun ultimately explode in supernovae, after they have exhausted their fuel, and stellar explosions can occur in another way as well. But a sign in the museum assured us, "It is more reasonable to believe that God created every star with its own individual characteristics, just as he has created billions of human beings, each one different from all others." Another statement on the wall provided further encouragement to the believers: "While the evolutionary view can interpret the evidence with some success, the Creationist Interpretation is always better. This is not surprising, for this is what the Bible says."

Why were the young visitors spared the scientific account of stellar evolution? Unfortunately, this explanation insists that the lifetimes of stars can extend over millions or billions of years, far beyond the thousands that arise from any literal interpretation of the Bible. In the view of Creation Science, these theories must then be incorrect.

Other inconvenient ideas were treated in a similar fashion in this museum, but no defense of Aristotle's spheres was presented. The Bible did not dictate stellar distances or positions, so modern astronomical estimates can be accepted, with some reservations. The Scriptures are even credited with predicting the huge population of stars that exist ("the host of heaven cannot be numbered," Jeremiah 33:22). As only 4,000 of these stars can be seen with the naked eye, the discovery of much greater numbers of them by telescope is cited as validation of the Bible.

The Creationists accept the numbers of the stars but have problems with their distances. If the universe is only 10,000 years old, then we should only be able to see stars 10,000 light-years away or less, which excludes most of our galaxy. Furthermore, stars should be winking on in the heavens continually as their light reaches us for the first time. A poster in the Museum of Creation and Earth History dismisses this difficulty. Either the light was put into position en route, during the Creation process, or the speed of light had a much larger value in the early days. Either way, the full heavens were made visible to us, perhaps to mark the seasons and aid the navigation of the seas.

As I continued my tour, I passed through rooms that could have passed muster in any science facility (for example, one with birds and fish). Others, however, strongly bore out the key Creationist belief in the literal correctness of the Bible. The unifying figure for their movement was a civil engineer, Henry M. Morris, who had come to the conclusion that "God doesn't lie." In his 1961 book with J. C. Whitcomb, Jr., *The Genesis Flood*, Morris argued that "the real issue is not the correctness of the interpretation of various details of the geological data, but simply what God has revealed in His Word concerning these matters." The museum had a suitable display, bathed in red light, to illustrate the Fall of Man, accompanied by comments that death had not existed before that event. The Second Law of Thermodynamics (we shall hear about its contents a bit further on) only started to operate after the Fall, as part of God's curse.

Another display showed the interior of Noah's Ark, with a long array of fenced-off compartments. The head of an ostrich peered out from one of them. The commands given to Noah by God (Genesis 6:19) had included these words: "And of every living thing of all flesh two of every *sort* shalt thou bring into the ark, to keep *them* alive with thee; they shall be male and female" (emphasis in original).

Calculations were displayed that stressed that the dimensions of the Ark (about $140 \times 23 \times 15$ meters) were equivalent to 569 railroad stockcars, and quite up to the task. My thoughts immediately went to the dinosaurs; in fact, a scaly tail was shown projecting from one normal-sized compartment in the illustration. What I yearned for, though, was a painting that illustrated the accommodations of the tyrannosaur and brontosaurus pairs on the Ark. None was provided, but I did find instead a painting of a peaceful scene in which a brontosaurus waded in a lagoon while modern antelope grazed on a hillside nearby.

No matter how they were housed on the Ark, a literal interpretation of the Bible prescribes that dinosaurs survived the Flood. Presumably, they persisted after that and interacted with humans for some time. Creationist literature accepts this possibility and equates dinosaurs with the dragons of mythology and the large beasts, behemoths, and leviathans described in the Bible. In an Institute for Creation Research publication, Henry Morris writes,

> Suppose the dinosaurs continued to survive for a time in the post-flood world. That would account perfectly for all the dragon stories, many embellished over the centuries with legendary accretions, but at the same time based on a substantial residuum of fact. . . . Most creationists believe that dinosaurs have co-existed with man from the beginning, only becoming extinct in the Middle Ages.

A 1993 article in *Creation* magazine was willing to escalate Morris's claims. It asked, "But could real dinosaurs be living today? What about all the reported sightings?" The article asserted that "fresh, unfossilized dinosaur bones have been found," though it later conceded that "the feasibility of the idea that some dinosaurs may still be alive has a little more support, although at this time we would have to say it is not conclusive."

The fictional display of dinosaurs in the film *Jurassic Park* evoked a different response from Morris, however. In his view, it was part of a dinosaur mania propagated by evolutionists, New Age cults, and Satanists. The special goal of these groups was to achieve "a world religion of evolutionary humanistic pantheism, with monotheistic religions, especially Biblical Christianity, banished from the earth."

During my visit to the museum, I learned that my presence in the museum had been noticed, and I was invited to meet the museum's senior scientist, Dr. Duane Gish. Dr. Gish is a Berkeley-trained Ph.D. biochemist who had worked in laboratories at Cornell University Medical College and the Upjohn Company before joining the Institute for Creation Research as Senior Vice President. His task in this last-named job was very different from those in his earlier work, however.

In his book *Evolution: The Fossils Say No!*, Gish demolished the purpose implied by the title of his Institute: "We do not know how the Creator created, what processes He used, *for He used processes which are not now operating anywhere in the natural universe.* That is why we refer to creation as special creation. We cannot discover by scientific investigation anything about the creative processes used by the Creator."

To what ends, then, does Dr. Gish use his scientific training? He seeks primarily to discredit those findings that are inconvenient for believers in the literal truth of the Bible. He has performed impressively in scores of debates against conventional scientists, in which his detailed knowledge of every wrinkle in their published data has worked to his advantage. The scientists, who had spent their time laboriously gathering the data rather than sharpening their debating skills, were sitting ducks, as Dr. Gish is quite adept in confrontations. An admiring colleague had commented that Gish "hits the floor, running," just like a bulldog, while Gish has added "I go for the jugular vein."

I didn't know what reception to expect as I entered Dr. Gish's office. I saw a graying, clean-cut, bespectacled, stocky, and rugged-looking man. My book *Origins: A Skeptic's Guide to the Creation of Life on Earth* was prominently displayed on the table. In it, I had criticized existing scientific theories on the origin of life, but I had hardly been kind to the Creationists. But I learned quickly that the credits I had earned for the former position easily outweighed my debits from the latter one. Dr. Gish and a colleague were cordial and appreciative of my role. They had confronted scores of critics, and the addition of another one mattered little to them. On the other hand, it delighted them to have a conventional scientist support them in one area of their own claims. We agreed that no satisfying scientific account of the origin of life exists and that existing attempts to fill that gap were seriously flawed.

I can understand the Creationists' delight at finding themselves on solid ground in at least one disputed area. In attempting to defend the literal truth of the Bible, they are often forced into ludicrous positions. They are compelled to attack techniques such as radioactive dating that are as well grounded as anything in science, and their attempts to do so discredit them thoroughly. How refreshing it must be for them to act as critics in a field where the positions are reversed and they are the ones who hold the better ground (I will elaborate on this in the next chapter).

After our happy beginning, I was treated to a long presentation by Dr. Gish concerning the gaps in the fossil record. This is not my area of expertise; Stephen Jay Gould was needed here. I could only comment that whatever the status of these gaps, they paled before the gap in the fossil record of Creation Science. If dinosaurs had coexisted with modern species, there should be abundant evidence of it: a tyrannosaur with a zebra in its jaws, for example. If the geologic ages all existed together, why not find traditions describing how our ancestors prepared a stew of trilobites (a predominant Cambrian creature)? I received no reply to this question, and Dr. Gish moved on to another area that may represent the Creationists' real agenda for action. I will paraphrase the question: If belief in God and the Bible is undermined by the teaching of contrary ideas such as evolution, what will be the consequences for society? *The Creationists* have left no doubt about how they feel on this. In *The Bible Has the Answer,* Henry Morris and Martin Clark have written that "evolution is not only anti-biblical and anti-Christian, but it is utterly unscientific and impossible as well. But it has served effectively as the pseudo-scientific basis of atheism, agnosticism, socialism, fascism, and numerous false and dangerous philosophies over the past century."

Another group, the Creation Science Research Center, has embellished this theme, claiming that evolution promotes "the moral decay of spiritual values which contributes to the destruction of mental health and . . . [the prevalence of] divorce, abortion and rampant venereal disease."

These points had not been neglected by the Museum of Creation and Earth History. A picture of Adolf Hitler was prominently displayed in the last hall, presumably as an example of the way in which evolutionary beliefs can lead us to evil. Before I departed, I suggested to Dr. Gish's assistant that the photograph be removed. A visitor passing through the hall in haste might misinterpret which side the Führer was supposed to be on.

The greater message could not be shrugged off so readily. Unlike the medieval cosmos, the new scientific view of the universe did not come equipped with any story or point of view to give it meaning, to lend significance to our existence. In effect, the Creationists' strategy was to stick with a previous vision that had a good track record for endurance and to trash any science that interfered with it. They are not an isolated fringe group.

A Gallup poll found that 44 percent of the American public endorses the following idea: "God created man pretty much in his present form at one time within the last 10,000 years." This viewpoint is obviously alive, though it suffers from the handicap that the observation of nature continually contradicts the believers' dogma. It may not be as easy for the faithful to do radioactive dating of rocks as it is for them to peer through the telescope, but nonetheless, the facts do tend to come up again and again.

The major religions of the United States keep their point of view much more in harmony with science. This was emphasized late in 1996 when Pope John Paul II accepted, on behalf of the Catholic Church, the likelihood of evolution. (An earlier Pope had declared that the theory was compatible with Church doctrine but did not declare it likely to be correct.) Pope John Paul did affirm the existing view that the spiritual soul is created by God. Many other faiths accept a similar position, and the Gallup poll that I cited previously listed the statement "Man has developed over millions of years from less advanced forms of life, but God guided this process, including man's creation" and reported that this statement was supported by 38 percent of the U.S. population. These religions have removed any profound contradiction between their teachings and the modern scientific model of the universe, but a deep aesthetic flaw is left in its place.

The major U.S. religions no longer deny the enormous size of the universe, but no importance is attached to it. The cosmos remains a decoration, just as it did when it was restricted to a set of nested celestial spheres. No role is given either to the billions of years that elapsed, more than half the lifetime of the universe, before the Earth or Sun existed. Perhaps some fast-forwarding mechanism ran them quickly by.

A few hundred years ago, these questions were considered more seriously, and efforts were made to fashion a view in which belief in a Creator was combined with acceptance of a larger universe. Unfortunately, the advocates of this vision made incorrect assumptions about the nature of the planets, and these mistakes brought their theories down. But we shall find it worth our while to revisit such theories, as some fragments of these assumptions still obscure matters today.

A Walk in the Garden

We will jump to a garden somewhere in France, on a clear evening in the late seventeenth century. Two lovers are strolling through the foliage and looking up at the starry heavens. One, a Parisian philosopher, is narrating the lovers' conversation. His companion is a noblewoman, a Marquise, who is eager to learn about celestial matters. She understands that the universe is larger than previously recognized, and she wants to know whether it is inhabited. The philosopher couches his reply in the language of love:

> Madam, *said I*, since we are in the humour of mingling amorous follies with our most serious Discourses, I must tell you that in Love and the Mathematicks People reason alike: Allow never so much to a Lover, yet presently after you must grant him more; nay more and more, which will at last go a great way: In like manner, grant but a Mathematician one little Principle, he immediately draws a consequence from it, to which you must necessarily assent.... Now this way of arguing have I made use of. The Moon, *say I*, is inhabited, because she is like the Earth; and the other planets are inhabited, because they are like the Moon. I find the fix'd stars to be like our Sun, therefore I attribute to them what is proper to that: You are now gone too far to be able to retreat, therefore you must go forward with a good grace.

The Marquise has caught his thrust: Since our own Sun illuminates its planets, every star must perform the same service. She is then rewarded with further details, some of them obviously tongue in cheek. An inhabitant of the Earth will obviously resemble a Moon dweller much more than a Saturnian. The worlds within the Milky Way approach one another so closely that birds may fly from one to the other and pigeons carry letters across.

Behind this wit lay a philosophy shared by many others: the idea that the Absolute Being, having created all that space, would not waste it irrationally but would use it to good purpose. The philosopher has based his argument strongly on analogy and what seems reasonable to him.

The philosopher points out that the planets resemble the Earth, which we know to be inhabited. The ones farther from the Sun are abundantly supplied with moons, to afford the inhabitants additional light. Furthermore, it is impossible to conceive of any other use for the planets, save as residences. All of this testifies to the magnificence and fruitfulness of nature.

The Marquise is not entirely pleased by these revelations: "We must confess," she comments, "that we scarce know where we are, in the midst of so many worlds; for my own part, I begin to see the earth so fearfully little, that from henceforth I shall never be concerned for anything . . . when anyone approaches me for carelessness, I will answer, *Ah, did you but know what the fixed stars are!*"

Her lover, of course, has his own view of things: "Dreadful, Madam, *said I;* I think it is very pleasant. When the heavens were a little blue arch, stuck with stars, methought the universe was too strait and close; I was almost stilled for want of air; but now it is enlarged in height and breadth . . . I begin to breathe with more freedom, and I think the universe to be incomparably more magnificent than it was before."

I have presented material from a book by Bernard de Fontenelle that was first published in 1686. If there had been a best-seller list at that time, this work would have had a prolonged run on it, as it went through many editions and translations from its original French into other languages. Its English title was *Conversations on the Plurality of Worlds.* The book undoubtedly owed its popularity to the romantic format in which it presented the material. The ideas that it presented had been voiced before and would be echoed for several centuries to come.

More than two centuries earlier, in 1440, Cardinal Nicholas of Cusa struck the same chord in his book *Of Learned Ignorance:*

> Life, as it exists here on earth in the form of men, animals and plants, is to be found, let us suppose, in a higher form in the solar and stellar regions. Rather than think that so many stars and parts of the heavens are uninhabited and that this earth of ours alone is peopled—and that with beings perhaps of an inferior type—we will suppose that in every region there are inhabitants, differing in rank and all owing their origin to God, who is the centre and circumference of all stellar regions. . . .
>
> It may be conjectured that in the area of the sun there exist solar beings, bright and enlightened intellectual denizens, and by nature more spiritual than such as may inhabit the moon—who are possibly lunatics—whilst those on earth are more gross and material.

Remarkably, these speculations preceded astronomical insights into the size of the universe.

By the eighteenth century, this point of view had become widespread, adopted by philosophers, scientists, and much of the general public. A new theology had been embraced in which God reigned in a universe populated by rational creatures. In the words of philosopher Arthur O. Lovejoy, a "great Chain of Being" existed, which aligned all creatures in an order from the most meager one to the highest possible kind of being, with humans somewhere in the middle. Lovejoy has written of this era, "there has been no period in which writers of all sorts—men of science and philosophers, poets and popular essayists, deists and orthodox divines—talked so much about the Chain of Being."

To sample this fare, we can note that Benjamin Franklin wrote in the *Poor Richard's Almanac* issue of September 1749 that "it is the opinion of all the modern philosophers and mathematicians that the planets are habitable worlds." He went on to discuss "superior beings in better worlds" who "smile at our theories and our presumption in making them."

Sir William Herschel, the astronomer who discovered the planet Uranus and believed the Moon to be inhabited, did not stop at that point. He supposed that the Sun was "most probably inhabited, like the rest of the planets, by beings whose organs are adapted to the peculiar circumstances of that vast globe." He was not considering the body we now know, with a temperature that ranges from thousands of degrees on the surface to millions in the center, but one whose interior was cooler than its outside.

Not everyone was happy with these amended versions of Creation. They posed problems for Christian theology in particular, which have persisted to this day. Philosopher Paul Davies expressed the dilemma recently: "Christianity faces a peculiar problem with regard to the Incarnation. Was this event unique in the universe, as official doctrine insists, or did God take on alien flesh, too? Is Christ the Saviour of humans alone, or of all intelligent beings in our galaxy and beyond?"

Others had considered the same dilemma and used it as an opportunity to reject the Christian religion. Thomas Paine (1737–1809), the American writer and advocate of the Revolution, wrote the following in *The Age of Reason* in 1793:

> Though it is not a direct article of the Christian system that this world that we inhabit is the whole of the habitable creation, yet it is so

worked up therewith from what is called the Mosaic account of the creation, the story of Eve and the apple, and the counterpart of that story—the death of the Son of God, that to believe otherwise, that is, to believe that God created a plurality of worlds at least as numerous as what we call stars, renders the Christian system of faith at once little and ridiculous and scatters it in the mind like feathers in the air. The two beliefs cannot be held together in the same mind; and he who thinks that he believes in both has thought but little of either.

The confrontation that finally developed in the nineteenth century over the existence of extraterrestrials took place between Christians, however. The advocates of a multiplicity of inhabited worlds did not base their opinions on telescopic observations. They deduced by reason how the universe must be and interpreted the evidence in support of their fixed positions. Their visions remain as an example of the practice of instructing nature rather than learning from it.

Nature, however, had a different picture to present and did so whenever it was given the opportunity. We have seen how the Moon did not give a very Earth-like appearance when inspected closely by telescopes. Observation of other worlds showed that each had its own character, but none resembled our own. A culminating debate took place in the 1850s between two prominent scientists, each of them devout believers.

The Great Extraterrestrial Debate

The dissenter to the concept of a universe of inhabited worlds was William Whewell (1804–1866), a multifaceted scientist who had written on tidal theory, astronomy, physics, inductive reasoning, and the history of science. He was, in fact, the first person to use the term *scientist:* In 1840 he wrote, "We need very much a name to describe a cultivator of science. I should incline to call him a Scientist."

At the time of the debate, Whewell was Master of Trinity College in Cambridge, England. His arguments were compiled in a book, *Of the Plurality of Worlds: An Essay,* which he attempted to publish anonymously, followed by a sequel, *Dialogue on the Plurality of Worlds.* He wrote to a friend concerning the first book: "I must publish my book without my name, in consequence of the heresies which it will thus contain." The effort to conceal his identity did not work.

Whewell's "heresy" was to argue that the Earth was likely to be the only place in the universe that held intelligent beings. Lower forms of life did not concern him. In reaching this conclusion, he drew upon examples from geology, astronomy, and theology. Much of his attention centered on our solar system:

> The Earth is really the only domestic hearth of this Solar System, adjusted between the hot and fiery haze on one side, the cold and watery vapour on the other. This region alone is fit to be a domestic hearth, a seat of habitation; and in this globe, by a series of creative operations . . . have been established, in succession, plants, animals and man.

Whewell saw, quite correctly, how little the other worlds resembled the Earth. With less insight, he also attempted to dismiss extrasolar planets, arguing that nebulae were made of gas and that double stars were unlikely to have planets. He had no difficulty with the idea that followed from his analysis: only Earth held intelligence. Historian Michael Crowe has summed up Whewell's position well:

> [The empty] planets and stars may be explained not by God having created them for living beings but as resulting from a general plan of creation of which the most noteworthy result is our inhabited planet. To the argument that if the other planets are not inhabited, then God has created them in vain, Whewell is now able to respond that God, because he works in general patterns, frequently appears to have worked in vain.

Whewell pointed out that few vegetable seeds become plants or animal ova, animals. He went on to emphasize that "one such fertile result as the Earth, with all its hosts of plants and animals, and especially with Man . . . is a worthy and sufficient produce . . . of all the Universal scheme." Reasserting the central Christian theme, he declared that "one school of moral discipline, one theatre of moral action, one arena of moral contests for the highest prizes, is a sufficient center for innumerable hosts of stars and planets, globes of fire and earth, water and air, whether or not tenanted by corals and madrepores, fishes and creeping things."

This situation, of course, resolves any problems concerning Christ's unique sacrifice on this planet. If only the Earth is inhabited, no conflict exists. Whewell observed the same alternatives as Thomas Paine but selected the opposite choice. His book *Plurality* went through five editions in England and two in America.

Whewell's opponent in this debate-in-books was Sir David Brewster (1781–1868), a Scottish physicist with special expertise in optics. During his lengthy career, Brewster invented the kaleidoscope, published more than 300 papers, and received several medals from the Royal Society. In spite of many scientific accomplishments, his reaction to Whewell's book was more emotional than analytic.

In an 1854 comment in the *North British Review*, Brewster expressed the opinion that opposition to life beyond this planet was an idea "which could only be regarded by an ill-educated and ill-regulated mind,—a mind without faith and without hope . . . a mind dead to feeling and shorn of reason." Whewell's geological argument was an "inconceivable absurdity which no sane mind can cherish, but one panting for notoriety," one "too ridiculous for even a writer of romance," and his opinions on the stars and planets "degrading to astronomy and subversive to the grandest truths." Another argument by Whewell was "the most shallow piece of sophistry which we have ever encountered in modern dialectics."

Brewster went on to publish a book in rebuttal, *More Worlds Than One: The Creed of the Philosopher and the Hope of the Christian* (1854), in which he, like Whewell, combined scientific and theological arguments. In contrast to Whewell, Brewster argued that all worlds must be inhabited, restating the common view of the previous century. A few quotes will serve to carry his drift:

That God would create a celestial body of significant size without bestowing life on it was both inconceivable and an impious idea.

Every planet and satellite in the Solar system must have an atmosphere.

The size or bulk of Jupiter is about 1300 times greater than that of Earth, and this alone is a proof that it must have been made for some *grand* and *useful* purpose. . . . [That planet may contain] a type of reason of which the intellect of Newton is the lowest degree.

Can we doubt, then, that every *single* star . . . is the center of a planetary system like our own. . . ?

The uniqueness of Christ's atonement did not trouble Brewster. Its force "does not vary with any function of distance" so "why may it not have extended to them all—to the planetary races in the past . . . and to the planetary races in the future?"

With the perspective of almost a century and a half, we can see that Whewell won the debate concerning the worlds of the solar system. They resemble his description much more than Brewster's. To our best knowledge, none of the solar system's worlds harbor intelligent life. The questions concerning lower forms of life in the solar system and intelligence in the universe still remain open, however.

In the nineteenth century, Brewster was the clear winner. Whewell's arguments did not sway the outlook of his time. The poet Alfred Tennyson wrote, for example, that "it is to me anything but a satisfactory book. It is inconceivable that the whole Universe was merely created for us who live in this third rate planet of a third rate sun."

Brewster also won the battle in the media. He outlived Whewell, and his book had a larger readership, selling 14,000 copies. It was republished for decades. But time has made the quarrel between Whewell and Brewster one of historical interest. Both ascribed the creative force in the universe to God. They only disagreed on the details of His plan. In the same century, however, further scientific developments were adding a very different point of view to the general debate.

The Decaying Universe

In 1856, when the Whewell-Brewster debate was still echoing in England, a German scientist was making "probably the most depressing prediction in the history of science." Philosopher Paul Davies has awarded this distinction to the claim that the universe was dying of a heat death. In more familiar terms, we might say that it was running out of gas. The father of this idea, according to Davies, was Hermann von Helmholtz (1821–1894), an incredibly prolific man who excelled as a physiologist, physicist, and physician (he had an affinity for professions that started with "physi").

A theme in von Helmholtz's work was the rejection of "Nature philosophy," as formulated by the German philosopher Immanuel Kant in the 1780s. Kant maintained that the mind, reflecting divine reason, could deduce the laws that governed the world from a few basic principles.

Helmholtz chose to learn from nature rather than deduce what must be there through philosophy.

The credit or guilt for the prediction, in my opinion, should be spread over a larger group of scientists, who collectively formulated and expanded the ideas called the Laws of Thermodynamics, in particular the Second Law. These concepts are central to the thread of our argument, so we must pause here to sample their flavor.

We will use a specific example. Yesterday evening, I wanted to meet a friend for dinner. To get to his house, I needed to drive my car, which was sitting in front of my own residence. If I tried to start the car by pushing it, I am sure that I would have to work very hard to roll it just a few feet forward. Instead of doing this I just climbed inside, inserted a small metal key into the ignition, and turned that key. Seconds later, several tons of metal, plastic, and glass, with myself in it, were speeding down the road. The automobile had gained a lot of energy of motion.

The First Law of Thermodynamics states that energy is neither created nor destroyed (an exception exists, but it is not relevant here), so my car must have obtained its supply from another source. The slight turn of a key with my fingers certainly didn't do it. It rather released a spark, which allowed the gasoline in my fuel tank to combine with oxygen from the air. This combustion of gasoline, forming carbon dioxide and water, released chemical energy.

I could have released the same chemical energy by pouring the gasoline over my car and igniting it with a match. A lot of heat and some light would have been produced, and some secondary chemical reactions would unfortunately have taken place as well, such as the charring of my upholstery. But the automobile would not have sped down the road. To get it moving, we need to *couple* the release of chemical energy to movement of pistons, axles, and wheels. The engine was constructed to do that job.

I needed to use only a small amount of gas from the tank to accelerate my car to 48 kilometers per hour. To keep it at that speed, though, I had to keep feeding more gas from the tank into the engine. Otherwise, the car would have come gradually to a halt, as the friction between moving parts in the engine, and between the tires and the road, slowed it down. The energy of motion of the car was transformed into heat energy in the road, the car, and the air. If the car could have been moved, by some miracle, into the space between the galaxies, it would have traveled an enormous distance without slowing down appreciably (never mind what would have happened to me). On Earth, however, the car eventually would have run out of gas. I would have had to refill the fuel tank if I wanted to drive farther.

The same considerations apply to all of us. We must eat food to keep on living and moving. Our bodies contain very intricate coupling devices that allow us to convert the chemical energy of the food and oxygen of the air to other kinds of chemical energy, which we store within us. Our internally stored energy is converted to the motion of our limbs and the electrical activity of our brains as needed. But if we stop taking in food or are cut off from an oxygen supply, we die.

Virtually all of the energy in our food and in gasoline originates in the Sun, which has been pouring that energy into space for almost 5 billion years. As I mentioned earlier, the Sun will run out of its own internal fuel supply in another 5 billion years or so. To survive beyond that, our descendants will have to switch to another energy supplier. How long will the stars last in general? Although the rate of star formation is slowing down, and the galaxies are gradually getting darker, new stars will continue to be made for a long time. On the time scale that I adopted for Creation Week, with one day equal to 2 billion years, the Sun will be gone in three "days," but future stars of its kind will continue to burn for 100,000 Creation "years" (10^{14} normal years). Other objects, such as red, white, and brown dwarf stars and black holes, will remain after that, lasting almost inconceivably longer times.

To picture this next time interval, let us create a new scale for time. On it, we will compress all of the time from the Big Bang until the period when the last of the normal stars disappears into a single second. The next era would last more than 10^{20} years on that scale.

Despite the incredibly long durations that are involved, all of the fuel must run out in the end, and all of the bodies named previously will disappear. Where then will the abundant stellar energy have gone?

The Second Law of Thermodynamics specifies that in every transaction, some of the energy is converted to heat, a form of energy that cannot be reused again. A cosmic sales tax drains everything we do. When an object heats up (my car, for example, if I had burned the gasoline on it rather than in the engine), the atoms and molecules in it move more rapidly, but at random. By contrast, when my car took off down the road, the atoms and molecules all moved together in the same direction, at the same rate. This contrast between concerted movement and randomness brings up an alternative way of expressing the Second Law.

Verbal descriptions are only approximate. Sportscasters use a host of different styles in following the plays on a football field, but each of them reflects the same reality. The Second Law has also been described in various ways. One important statement of this law stresses that only those events can take place in the universe (or in any sealed-off part of it) that increase the total amount of disorder, or randomness.

Common sense tells us this is so, as we experience it too often in our lives. As I use my car, parts wear out or break. The window may get scratched, tires wear out or puncture, and brakes need relining. Even if I left the car unused, I would find that the parts gradually rusted. The reverse doesn't take place. Windows don't unscratch and tires don't repair themselves. That is not the way of the universe. The term that scientists prefer to use to describe this tendency to scatter, degrade, wear down, or wear out is *entropy*, and I will use it as well, together with its equivalents: disordered, disorganized, more random. We can then restate the Second Law as follows: Events may occur spontaneously, provided that the entropy of the universe increases as a result.

I would prefer a more poetic descriptor, one borrowed from the *Star Wars* film series: I would call entropy The Dark Side of the Force. As a result of the Dark Side's existence, all of the useful energy in the universe eventually will be converted to heat, and everything will then shut down.

A Sour Taste

For most of us, the eventual death of the universe, or even of the Sun, would not seem a crisis that we would place ahead of the next income tax payment on our personal list of emergencies. Problems lose their impact when we are given an inconceivable number of years in which to solve them. For philosophers, theologians, and others who speculate on what it all may mean, these concepts had a depressing effect. This information went hand in hand with the new observations concerning the other worlds of our solar system.

The other planets were not extensions of the Earth. An unprotected human being placed on any of them would die almost immediately. The majority of them offered, like the Moon, rocky surfaces exposed to the vacuum of outer space, baked by the Sun or frozen by the cold. The remainder had atmospheres of poisonous gases: some searingly hot, and others unbelievably frigid. The cosmos was not created for the comfort of human inhabitants.

When these revelations were combined, they provided a picture of an alien, disintegrating universe, which stood in bleak contrast to the earlier medieval one. This model was unlikely to have an inspirational effect on society. For an apt summation, I will borrow once more from the character of Cardinal Bellarmine in Barrie Stavis's play, *Lamp at Midnight:*

> What will happen to the masses of men who have been nurtured in the belief that the world was created for man, and that he is God's special concern? They would feel cheated, belittled, denigrated. They would turn in revulsion. Heresy, apostasy, atheism would be the order of the day. You would create a spiritual revolution.

The masses, to some extent, have resisted that fate by the expedient of ignoring the universe and getting on with their daily lives. But the impact on intellectuals has been profound. A selection of quotes will bring the point home. A choice starting point is a much cited passage from a 1903 essay of the British philosopher Bertrand Russell:

> Such, in outline, but even more purposeless, more void of meaning, is the world which science presents for our belief. Amid such a world, if anywhere, our ideals henceforward must find a home. That man is the product of causes which have no prevision of the end they were achieving; that his origin, his growth, his hopes and fears, his loves and his beliefs, are but the outcome of accidental collocations of atoms; that all the labors of the ages, all the devotion, all the noonday brightness of human genius, are destined to extinction in the vast death of the solar system, and that the whole temple of man's achievement must inevitably be buried beneath the debris of a universe in ruins—all these things, if not quite beyond dispute, are yet so nearly certain that no philosophy which rejects them can hope to stand. Only within the scaffolding of these truths, only on the firm foundation of unyielding despair, can the soul's habitation safely be built.

Decades later, Nobel Laureate biologist Jacques Monod embellished this theme in his influential 1971 book *Chance and Necessity.* Monod wrote of the legacy of early human societies: "From them we have probably inherited our need for an explanation, the profound disquiet which goads us to search out the meaning of existence. The same disquiet has created all the myths, all the religions, all the philosophies, and science itself." This need, Monod felt, may even be inscribed in our genes. Science, however, offered nothing more now than a vision of "a frozen universe of solitude." How then should this message be regarded? Monod had a prescription:

> If he accepts this message—accepts all it contains—then man must at last wake out of his millenary dream; and in doing so, wake to his total solitude, his fundamental isolation. Now does he at last realize that, like a gypsy, he lives on the boundary of an alien world. A world that is deaf to his music; just as indifferent to his hopes as it is to his suffering or crimes.

A brief, but potent, variation on this theme was produced by Nobel Laureate physicist Steven Weinberg near the end of his cosmological account *The First Three Minutes:*

> It is almost irresistible for humans to believe that we have some special relationship to the universe, that human life is not just a more-or-less farcical outcome of a chain of accidents reaching back to the first three minutes, that we were somehow built in from the beginning. . . . It is hard to believe that all this [Earth] is just a tiny part of an overwhelmingly hostile universe. . . . The more the universe seems comprehensible, the more it also seems pointless.

Many more quotes could be provided. The authors include many whose work in science, or in the explanation of science to the public, I admire greatly. Their viewpoint has spread and pervades much of the intellectual discussion of this century. It will accompany us as we move through this book. For our convenience, we will want to find a simple name to refer to it, because none is in common use.

To start, we have to recognize first that this philosophy is not just an impartial description of the universe, as we understand it today through science. It holds a lot of emotional content, which is not required by the science but conveys the personal positions of the writers. These positions bring up a memory from my adolescence. I had a friend who, whenever we came upon a group of people with particularly unpleasant expressions on their faces, would comment, "They look like they've been sucking sour lemons." That observation will serve us quite well now. I shall call the group of philosophers that I have quoted the Sour Lemon School.

The name expresses my reaction to their emotional tone; I do not intend to quarrel with the scientific content that supports their philosophy. But the science does not mandate that the universe is hostile or pointless, or that we are doomed to extinction. We have a lot to say about these matters, and it will be our actions that determine whether or not the members of the Sour Lemon School are right.

The Sour Lemon School emerged from an intellectual revulsion against the attempts to preserve the medieval cosmos by dismissing the discoveries of science. But it goes further to maintain that no new story, consistent with science, can be constructed. In the words of Leo Tolstoy, "The meaningless absurdity of life is the only incontestable knowledge accessible to man."

Evolutionist Richard Dawkins has updated and embellished this theme:

> In a universe of electrons and selfish genes, blind physical forces and genetic replication, some people are going to get hurt, other people are going to get lucky, and you won't find any rhyme or reason in it, or any justice. The universe that we observe has precisely the properties we would expect if there is, at bottom, no design, no purpose, no evil and no good, nothing but pitiless indifference.

Only one blemish exists on this diorama of a uniformly bleak and pointless universe that, like a pendulum, is gradually slowing to a halt. That exception surrounds us on all sides.

In this world beautiful flowers bloom outside my door in the spring, radiant colors appear in the autumn, and lovers still walk hand in hand in the gardens. I can bring up the music of Mozart or Beethoven in my own home when I choose, or get on the Internet and follow the progress of a spaceship as it circles around Jupiter. Some of these wonders existed long before humans appeared; others have come up only recently. These events do not fit readily into the picture of a universe that is relentlessly running down, yet they have taken place within the framework of the Second Law.

Nobel Laureate Christian De Duve wrote, in response to the comment of Steven Weinberg, "I view this universe not as a 'cosmic joke,' but as a meaningful entity—made in such a way to generate life and mind, bound to give birth to thinking beings able to discern truth, apprehend beauty, feel love, yearn after goodness, define evil, experience mystery."

To complete their philosophy, followers of the Sour Lemon School must explain how these aberrations infiltrated their cosmos. I will again call on Jacques Monod to express their point of view (the italics are his):

> Chance *alone* is at the source of every innovation, of all creation in the biosphere. Pure chance, absolutely free but blind, is at the very root of the stupendous edifice of evolution.
>
> . . . Life appeared on earth: what, *before the event*, were the chances that this would occur? The present structure of the biosphere far from excludes the possibility that the decisive event occurred only once. Which would mean its *a priori* probability was virtually zero.
>
> This idea is distasteful to many scientists. Science can neither say nor do anything about a unique occurrence.
>
> If it was unique, as may perhaps have been the appearance of life itself, then before it did appear its chances of doing so were infinitely slender. The universe was not pregnant with life nor the biosphere with man. Our number came up in the Monte Carlo game.

In the preceding selection, Monod appears to recognize that reliance on extreme chance represents the weak point of the philosophy. In science, chance is used as a last measure when all efforts to discover some deeper law have failed. We shall see that in the case of life, the efforts have hardly begun. But first another point requires attention.

Humans appear to need some vision, some context on which to build their dreams and plan their actions. Philosophies and religions usually contain within them some prescription for human behavior. How does the Sour Lemon School recommend that we proceed?

Bertrand Russell, in his essay cited previously, tried to construct an attitude: "To defy with Promethian constancy a hostile universe, to keep its evil always in view, always actively hated, to refuse no pain that the malice of power can invent, appears to be the duty of all who will not bow before the inevitable." But in the same work, Russell assured us that such defiance would be futile: "Brief and powerless is man's life; on him and all his race the slow, sure doom falls pitiless and dark. Blind to good and evil, reckless of destruction, omnipotent matter rolls on its relentless way."

Jacques Monod also tried to blaze a path: "The ancient covenant is in pieces; man knows at last that he is alone in the universe's unfeeling immensity, out of which he emerged only by chance. His destiny is nowhere spelled out, nor is his duty. The kingdom above or the darkness below: It is for him to choose."

But what should be chosen? Monod advocates the quest for knowledge, humanism, and other virtues and hopes that they may calm the fear of solitude and satisfy the need for an explanation. He is very conservative in judging the possibility of success, however, concluding only that "it may not be altogether impossible."

These spokesmen have been most eloquent, but I have seen no rush of volunteers eager to construct an encouraging view of the human future on a foundation of despair. The dilemma of modern humans has been summarized most eloquently by the physician and gifted science popularizer Lewis Thomas:

> We are in trouble whenever persuaded that we know everything. Today, an intellectually fashionable view of man's place in nature is that it makes no sense, no sense at all. The universe is meaningless for human beings: we bumbled our way into the place by a series of random and senseless biological accidents. The sky is not blue: this is an optical illusion—the sky is black. You can walk on the moon if you feel like it, but there is nothing to do there except look at the earth, and when you've seen one earth, you've seen them all. . . .
>
> I cannot make my peace with the randomness doctrine: I cannot abide the notion of purposelessness and blind chance in nature. And yet I do not know what to put in its place for the quieting of my mind. . . . We talk—some of us, anyway—about the absurdity of the human situation, but we do this because we do not know how we fit in, or what we are for. The stories we used to make up to explain ourselves do not make sense anymore, and we have run out of new stories, at least for the moment.

A New Story: First Draft

The time is ripe for a new vision. If we seek one that accepts the scientific picture of the cosmos but casts it in a positive light, then we must reverse the position of the Sour Lemon School. The existence of life, which is their anomaly, serves as our point of inspiration. We start with the assumption that life arose through processes that do not depend on extreme chance, but are built into the laws of the universe. These laws have operated since the Big Bang and governed the successive appearance of galaxies, stars and planets, life, and finally consciousness. The name most often applied to the overall process is Cosmic Evolution.

At first glance, these evolutionary processes appear to contradict the Second Law of Thermodynamics. The organization that exists in the human body and the information that we have stored in libraries and computers represent extraordinary concentrations of negative entropy—the reverse of randomness and disorder. How could these arise in a universe where every transaction must be paid for by a gain in entropy? A loophole exists.

Entropy must increase for the entire universe, but a huge local deficit is permitted, provided it is balanced by a greater gain in entropy somewhere else. In the case of life on Earth and all of its products, the increase in organization is "paid for," ultimately, by the continuing increase in the entropy of the Sun. One analogy might be to compare us to a charity whose works lose money but that is sustained by outside philanthropy. In the view of cosmic evolution, this development of local concentrations of increasing negative entropy is expected, given the laws of our universe. We could think of it (again with apologies to George Lucas) as the Light Side of the Force, or just the Force.

A compact, if more formal, description of the process of Cosmic Evolution can be found in an article by biochemist John Keosian, published in 1974:

> Matter driven by energy in an open system can go to higher and higher levels of organization. The thing to bear in mind is that each level of organization has its own properties by which it alone can best be recognized. Also, each higher level, although incorporating structures and processes evolved at the lower levels, has new properties not predictable from the properties of the lower level. This is true of the whole progression from elementary particles through atoms through molecules to man. Each stage in that progression incorporates structures and processes of the lower level but emerges as a new stage with new properties and the propensity for arriving at a higher level of organization.

These ideas have percolated through the scientific establishment, filled courses on college campuses, and emerged in NASA's new Origins program, which asks, "How have order and structure emerged in the cosmos?" But such ideas have never quite impacted the public through the media. One coherent statement of their potential was made by Eric Chaisson, director of the Wright Center for Science Education at Tufts University:

> The scenario of cosmic evolution will give us an opportunity to systematically inquire into the nature of our existence. As we approach the end of the millenium, such a coherent story of our origins—a powerful and true myth—can act as an effective intellectual vehicle to invite all our citizens to become participants, not just spectators, in the building of a whole new legacy.

More recently, NASA has established an Astrobiology Institute to fund research in this area. Efforts at eleven separate organizations, including, for example, Harvard University, the Scripps Research Institute, and the Jet Propulsion Laboratory, were selected for support in the initial competition in 1998. According to NASA's Gerald Soffen, the interim coordinator of the institute, astrobiology addresses the question, "Is life a cosmic imperative?"

One strength of the Cosmic Evolution viewpoint lies in its rationale for the great age of the universe. If we choose to value not just our own existence, but the entire process that gave rise to us, then the billions of years that preceded our appearance become a necessary prelude to our own history. Each useful new biochemical advance in evolution marked a triumph, even if no newspapers were around to celebrate it. The organisms that preceded us, such as the Cambrian chordate Pikaia, become legitimate parts of our own family tree.

The vast size of the universe also meshes well with the Cosmic Evolution story. We can view the entire terrain of Level 24 as one vast incubator for life. We represent one of the successful experiments. If we value the process that has allowed life to spread out from its birthplace on this planet to occupy most of the surface of the globe, then we will want to carry the process further. We are ready to expand into the enormous playing field that surrounds us and see what else may have evolved out there. In that spirit, the endless stars and galaxies of the cosmos represent future sites for our exploration, and not just a waste of space.

Another motive can drive us as well. Nothing in the scheme I have described suggests that evolution need end with us in our present state. The history of our biosphere has been one of experiment and diversification. If we value the steps that have produced us, then we will want to take them further, in a host of directions. In doing this, we need not risk what we have now—we can have our cake and eat it, too. The vastness of the universe leaves a lot of room for experiment. We could leave humanity as it is, on this planet, and use others for exploring our further evolution.

A Fair Contest

Cosmic Evolution incorporates and accepts the findings of science but also involves several value judgments that are outside of science. It is opposite in its spirit to Sour Lemon teachings, and it differs from many traditional religions, though it is compatible with others. Is the choice between all of these philosophies just a question of individual preference? The answer is no, because the validity of Cosmic Evolution depends on an assumption that is testable. It presumes that self-organizing forces operate on a broad scale in nature. Sour Lemon advocates and Fundamentalists declare that no such forces exist. This issue could be contested on many fields, but the most crucial one, the weakest link in the Cosmic Evolution chain, is the problem of the origin of life.

Life exists on this planet. We do not know how it began, but three strong positions have been staked out. Philosopher Paul Davies has classified and discussed them in his book entitled *Are We Alone? Philosophical Implications of the Discovery of Extraterrestrial Life.*

1. Life began through a miracle. The Bible, of course, describes this explicitly. According to Davies, "Many people feel that if life in general, or the human species in particular, had a completely natural origin, it would undermine any claim to our occupying a special place in the scheme of things and break one of the most powerful bonds that religion has claimed exists between humans and God."
2. Life began through an accident, but one that did not violate the laws of physics. A series of natural events, each of which is of very low probability but lawful, would qualify as a scientific explanation, even if it appeared miraculous. This, of course, represents the viewpoint that I have called the Sour Lemon School.
3. Life began through natural processes of high probability. They represent part of the greater scheme of Cosmic Evolution.

For simplicity, I will use these classes as a basis for future discussion, but I must mention that each holds important subgroups. Fundamentalists maintain that God made the universe and life within seven days, using powers outside of scientific observation. Many other religions accept a miraculous origin but are willing to concede that it may have taken place billions of years ago.

Yet another subclass conceals the supernatural component by invoking the concept of Intelligent Design. Advocates of this process maintain that natural law is insufficient to generate the life we know from nonlife, unless an intelligent being were guiding the process. Presumably, this intelligence could be a preexisting life form of another kind, operating by natural law (we will encounter one such suggestion), and no miracle would be involved. But then one must account for the origin of the preexisting life form. The existence of a Supreme Being as the ultimate Creator is implied in this scenario, even if He is not mentioned explicitly.

We cannot settle this argument concerning our origins directly. No evidence remains from the earliest days of the Earth. But the Sour Lemon and Cosmic Evolution scenarios make very different predictions about the likelihood of life starting independently on other worlds, and this question is one that we can decide by making observations. The planets await our exploration.

If life started more than once in the same solar system, then a general process would appear to be at work rather than an enormous stroke of luck. If we can collect enough information to understand the process and place it within the framework of science, then no miracle would be needed for our explanation, though a Fundamentalist could still hold that one was involved. For example, I asked a guide at the Museum of Creation and Earth History how he would react if life were discovered on Mars or Europa. "If it exists," he replied, " the Creationists would hold that God created it."

On the other hand, if a thorough search of our system turned up no evidence for life, or even for a more limited self-organization process, then Cosmic Evolution would be profoundly weakened. The defeat would be most stinging if a suitable environment was discovered (for example, an ocean well equipped with suitable energy sources) and it turned out to be utterly barren. Theories invoking a miraculous creation or an unlikely stroke of luck would be strengthened, in that both view the universe as hostile to life and normally barren.

The contest has not really begun, and the outcome is wide open.

Whatever we may find, we will truly be surprised. But before we plan where to search, we find out why the origin of life question poses such a severe scientific problem. This difficulty is relatively new. For much of human history, the origin of simple creatures was thought to take place rather easily. In the next chapter we learn of the discoveries that triggered the change of perspective.

4

Life in the Museum

L ife flourishes on this planet, and we take it for granted. We are used to the fact that it may turn up exactly when it is least wanted: as mold on the bread and cockroaches in the kitchen. At some point we may wonder why these creatures turn up so readily. Believers in the medieval cosmos accepted the idea that the various forms had been created by God during Creation Week, but they did not think that each new and unexpected appearance had to be credited to Him. Instead, another simple explanation crept in, that of spontaneous generation. This concept was challenged in the nineteenth century, at a time when many other traditional ideas were also under attack. Finally, a showdown was called between the leading protagonists, to settle the spontaneous generation question.

Is Paris Boiling?

The great boiled urine confrontation of 1877 never did take place, perhaps denying the world some excitement of the type usually reserved for sporting events. The challenger, Professor Henry Bastian of University College, London, certainly was courageous enough. He was willing to stage the contest in Paris, the home city of his opponent, the celebrated Louis Pasteur. Bastian also accepted an additional handicap. Two of the judges, politician Jean Baptiste Dumas and naturalist Henri Milne-Edwards, had been avid supporters of Pasteur in earlier phases of the controversy. The third, Phillippe van Tiegham, was a former pupil of Pasteur. Yet Bastian felt secure enough of his position that he was ready to demonstrate it and defend it in this arena.

Bastian had published results that appeared to confirm spontaneous generation. Professor Pasteur, after altering the conditions a bit, had obtained very different ones. Bastian wrote:

> Further discussion between M. Pasteur and myself seems to me in the present phase of the question to be almost useless. Certainly, no good can come from our alternate enunciation of opposite experimental results, when precisely the same methods have not been had recourse to. I am perfectly willing to reproduce before competent witnesses the results of which I have spoken.

Pasteur quickly accepted the contest: "I defy Dr. Bastian to obtain, in presence of competent judges, the result to which I have referred."

The theory of spontaneous generation, as set forth by biologist and historian John Farley, maintains that "some living entities may arise suddenly by chance from matter independently of any parent." This possibility had not been an issue through most of human history. It was accepted in most societies and endorsed by such noted thinkers as Aristotle, St. Augustine, Thomas Aquinas, Francis Bacon, Galileo, Copernicus, and Goethe. In Shakespeare's *Antony and Cleopatra* (Act II, Scene 7), Lepidus proclaims, "Your serpent of Egypt is bred now of your mud, by the operation of your sun; so is your crocodile."

In the sixteenth and seventeenth centuries, mere passive observation was replaced by specific recipes. For example, the following comes from the seventeenth-century Flemish biologist Jan Baptiste van Helmont:

> If you press a piece of underwear soiled with sweat together with some wheat in an open mouth jar, after about twenty-one days the odor changes and the ferment, coming out of the underwear and penetrating through the husks of the wheat, changes the wheat into mice. But what is more remarkable is that mice of both sexes emerge [from the wheat] and these mice successfully reproduce with mice born naturally from parents. . . . But what is even more remarkable is that the mice which came out of the wheat and underwear were not small mice, not even miniature adults or aborted mice, but adult mice emerge!

More remarkable yet is the absence of skepticism in the account. I suspect that the precautions needed to keep common house mice out of the wheat-and-underwear preparations should not be too difficult to put into effect. Later workers applied more common sense in their experimental designs and got negative results for a whole array of lower life forms. By the mid-nineteenth century the spontaneous generation question had contracted to the realm of microorganisms. As Farley put it,

> The cell theory was modified in the 1860's so that protoplasm came to be considered the structural unit of life. Moreover, this protoplasm was viewed as an essentially simple substance, theoretically capable of being produced directly from inorganic matter. That the simplest organisms were generally regarded merely as naked lumps of protoplasm added credence to the belief that they, too, could be produced spontaneously.

How then would you test the idea that a suitable broth could give rise, spontaneously, to living bacteria? The concept is simple: sterilize your broth by heating, to kill any bacteria that are present, seal it off to prevent contamination by living bacteria, and watch it. Unfortunately, different experimenters got different results, usually the ones that agreed with their prior prejudices. Some noteworthy quarrels arose as a result.

In the eighteenth century, John Turberville Needham, a Welsh naturalist and Jesuit priest, consistently observed the appearance of "animacules" in broths that he had boiled and sealed. He considered this a proof of the spontaneous generation of microbes from nonliving organic matter. Others disagreed. Lazzaro Spallanzani, an Italian scientist-priest, heated his preparations for a longer time, to ensure sterilization, and took extra precautions in sealing his containers. He observed no spontaneous generation. Needham, however, did not relent. We can quote his words:

> He [Spallanzani] hermetically sealed nineteen vessels filled with different vegetable substances and he boiled them thus closed, for the period of an hour. But from the method of treatment by which he had tortured his nineteen vegetable infusions, it is plain that he has greatly weakened, and perhaps entirely destroyed, the vegetative force of the infused substances.

Thus the heat that one side claimed was needed to destroy existing bacteria was taken by the other side to have destroyed the ingredients needed for spontaneous generation.

This impasse lasted for nearly a century, until the entrance of Louis Pasteur. His celebrated studies on bacterial fermentation had called other problems to his attention: those of the origin of bacteria and of the diseases caused by bacteria. He succeeded so well with an elegant series of experiments that in 1862 he was awarded a prize of 2,500 francs by the French Académie des Sciences. The award was intended for "him who by well conducted experiments, throws new light on the so-called question of spontaneous generation." In a lecture at the Sorbonne, Pasteur was able to proclaim, "never will the doctrine of spontaneous generation recover from the mortal blow of this simple experiment."

In one of the best-known and simplest of the experiments that supported his claim, Pasteur sterilized a broth of sugared yeast water by heating it. The flask was not sealed but remained in contact with the outside air through a long swan (S-shaped) neck. The shape of the flask protected the contents from the entry of dust, which remained trapped within the neck. The bacteria carried by the dust had no chance to enter the flask, and its contents remained sterile indefinitely. When the neck was removed, however, bacteria soon reappeared within the broth, demonstrating that no life-supporting force had been destroyed by the heat.

Despite the "mortal blow," supporters of spontaneous generation persisted. Prominent among them was Felix Pouchet, director of the Natural History Museum at Rouen. He insisted that he could obtain spontaneous generation from broths made of hay, and he contested many of Pasteur's findings. In the tradition of French science at that time, a Commission was appointed by the Académie des Sciences to resolve the dispute. It supported Pasteur, issuing "a report which scarcely veiled its contempt for Pouchet and his colleagues." The report was subsequently endorsed by the Académie.

Matters rested there, as far as the French were concerned, until the next decade, when Henry Bastian and others took up the question again. Like Pouchet, Bastian claimed to have obtained spontaneous generation with hay infusions and, above all, with boiled, carefully neutralized urine. Pasteur denied the claims, as before. Rather than have a Commission judge the issue on the basis of published reports, however, Bastian insisted that a live demonstration be performed before the Commission. The commissioners hesitated over this prospect, and there was lengthy wrangling over the ground rules. Finally a compromise was arranged between Bastian and J. B. Dumas, who was Secretary of the Académie as well as a Commission member. On the morning of July 18, 1877, Pasteur and Bastian presented themselves for the urine contest. The only items that were generated, however, were chaos and bad behavior. Dumas did not appear at first. Milne-Edwards and van Tiegham did show, but when the former learned of the rules (Dumas had not kept him informed), he refused to endorse them and left the site. Finally Dumas arrived, but when he learned that Milne-Edwards had left, he declared that the Commission was at an end, and he left as well. The game had been canceled but no rain checks were offered.

The commissioners had avoided the contest with good reason. Bastian might well have succeeded in his demonstration, as Pasteur's clever experiments had a serious flaw. The amount of heating that Pasteur applied in sterilizing his yeast extracts was suitable for his own preparations but insufficient to kill some resistant bacterial spores that may be found in hay and urine. The results of Pouchet and Bastian were valid, but their interpretation was incorrect. They had discovered heat-resistant microorganisms rather than spontaneous generation.

The way in which the contest was conducted suggests that much more was at stake than questions of bacterial birth and survival. For insight, we can turn again to Pasteur's Sorbonne lecture:

> Scientific controversies are much more lively and passionate now because they have their counterparts in public opinion, divided always between two great currents of ideas, as old as the world, which in our day are called materialism and spiritualism. What a victory for materialism if it should be affirmed that it rests on the established fact that matter organizes itself, takes on life itself, matter which has in it already all known forces! . . . Of what good would it be then to have recourse to the idea of a primordial creation, before which mystery it is necessary to bow? To what good then would be the idea of a creator God?

France had been polarized and in political conflict since the time of the French Revolution. Finally, some stability had been reached in the Second Empire of Louis Napoleon. The Church stood in support of that regime and felt that the alternative was atheism, revolution, republicanism, and scientific materialism. The recently published work of Charles Darwin was seen as additional encouragement for the opposition, and spontaneous gen-eration went hand in hand with evolution. In the words of John Farley, "To the French, the doctrine of spontaneous generation was part of a broad package of politically and religiously dangerous doctrines, and ought to bear its share of guilt for the horrors of the recent past."

Supporters of evolution and nontheists, on the other hand, felt that spontaneous generation was necessary to complete their position. In the words of botanist Karl Nägeli, "To deny spontaneous generation is to pro-claim a miracle."

Despite the implications, belief in the abrupt spontaneous generation of microorganisms slowly withered, and by the early twentieth century, Henry Bastian was the sole surviving advocate. It was not a round of additional sterilization experiments that dispatched the concept, but rather the grow-ing realization that bacteria, and the protoplasm they contained, were not simple. This contradicted a common scientific idea of the mid-nineteenth century—that the building material of life was a pulpy, gel-like, structureless material called protoplasm.

This material could be packaged within walls and might contain an extra unexplained feature called the nucleus, but protoplasm itself con-tained the essence of life. Among the advocates of this idea were zoologist Ernst Haeckel and evolutionist Thomas Henry Huxley. In 1857, Huxley had examined mud that he had dredged from the sea bottom and noted that it contained a "transparent gelatinous matter," which he concluded was pro-toplasm; "Urschleim" was an alternative name provided by Haeckel. Biolo-gist John Browning summarized the idea: "There is no boundary line between organic and inorganic substances . . . reasoning by analogy, I be-lieve that we shall before long find it an equally difficult task to draw a dis-tinction between the lowest forms of living matter and dead matter."

With these concepts in mind, it is not surprising that some scientists felt that protoplasm and bacteria could form readily from sugar water. To un-derstand how far off the mark they were, we shall make a short tour of bac-terial structure as biochemists understand it today.

The Lower Floors of the Museum

Again, we can be helped by a visit to our as yet imaginary Museum of the Cosmos. The lower floors will present constructed rooms that appear to shrink us, rather than swell us, in size. The floor or level marked −1, in keeping with our use of an exponential system, will represent a world in which we are one-tenth our present size. On floor -2, we will be one-hundredth of our normal dimensions, and so on. As those of us know who have seen such films as *Honey, I Shrunk the Kids* and *The Incredible Shrinking Man*, we can have a lot of fun designing the sets. But we are after a particular quarry, the simplest kind of living creature, so we will skip over those levels and go farther down to level -5. This is the appropriate stage for bacteria hunters. To enter their realm, we have had to shrink to a hundred-thousandth of our present size.

We will need a point of reference from our familiar lives, to help us get oriented. The dot from the letter "i" on this line will have expanded to a huge black spot the size of a baseball diamond (roughly 30 meters across). We will leave out the rest of the "i," as it would extend for more than two city blocks. Instead, we will fill the space with models of various micro-organisms. Many of us will remember the paramecium from high school biology classes; normally it lives in freshwater ponds. On the -5 level we would confront a 15-meter-long cigar-shaped monster that was covered with hundreds of hairlike filaments. If we looked more closely, we would notice a pore on one side, used for sweeping in food, and the presence of endless complexities within.

We can move on. The paramecium is constructed on the more elaborate eukaryotic cell plan; we are looking for something simpler. The most ancient microfossils had shapes that resembled those of prokaryotes, simple blue-green algae, and bacteria. The latter were also the desired products in Henry Bastian's spontaneous generation experiments. We have furnished our -5 level with a model of *Escherichia coli*, a simple bacterium that lives in our intestines.

Our model would resemble a rounded cylinder, about the size of my upper arm, with a number of longer slender filaments extending from it. If we chose to animate our model, we would have the filaments whip back and forth, moving the creature around the room. Between the filaments, we would find many short and stubby hairs, which cover the bacterial body. But our principal interest would lie inside.

A mid-nineteenth-century model builder would have packed the interior with a uniform gelatinous substance, to represent the protoplasm. With a hundred and fifty years of additional scientific study under our belts, we are obliged to include much more internal architecture. As we are building our model for educational purposes, we can take a few liberties to make the plan more understandable. To start, we will want to identify the different kinds of construction materials that are present. In our own buildings, we use metal, glass, stone, and wood; a bacterium is constructed of proteins, nucleic acids, carbohydrates, and lipids. Most of us can tell stone from glass, but we could not scan a bacterial model and distinguish one chemical from another. So we will borrow a device from some biochemistry texts and color code them, blue for proteins, pink for nucleic acids, green for carbohydrates, and yellow for lipids.

A single glance at our bacterial model informs us that it is very complex. The outer skin is made of three separate layers, and a hooklike device of rods and rings is used to connect each of the long filaments to the body. The inside looks like the mother board of a computer, or a part of a telephone exchange, as a network of wires and tiny beadlike structures fills it. In our model, blue and pink components dominate the interior, indicating two of the building materials, proteins and nucleic acids. Yellow predominates in the "skin," signaling the presence of lipids, and the short stubby hairs that project from the skin are colored green, for carbohydrates. The filaments are made of protein, so we again use blue. From the intricacy of this construction, we can see why the experiments of Henry Bastian and his colleagues were doomed. To borrow a comparison once used by astronomer Fred Hoyle, we would no more expect a mixture of simple chemicals to self-assemble into this structure in a few days than we would anticipate that a tornado, blowing through a junkyard, would build a Boeing 747 aircraft.

A brief inspection of a bacterium at the -5 level has convinced us of its intricacy but not provided us with any "feel" for what is going on inside. We need to probe more deeply, so we will try the -8 level. At that degree of enlargement, 10 million fold, our creature would approximate the size of an ocean liner in size and complexity and be as strange as an alien spaceship. We would need a full course in cell biology to explore the corridors and rooms of this vessel. For our purposes here, though, we can be satisfied with a sample. Imagine that a side room at this level contains just one of the roundish blue components that appeared as a bead at the -5 level.

Anatomy of a Protein

The color blue is very visible in the interior of our model as proteins are a favored building material for bacterial construction. Hundreds or thousands of different ones can be found, with a bewildering profusion of sizes and shapes, but only a single example stands before us now. We see an irregular squarish object with a cleft on one side that at the -8 level approximates the size of a large valise or beach ball. We will not learn too much by watching a static model, but our imaginations are not limited by a budget, so we can pretend that this model has been mechanized. To appreciate the action, though, we must understand what this protein does.

We have selected hexokinase, a member of the subclass called enzymes. We can think of the enzymes as small machines, tiny facilitators, that help things happen in a cell. The job of hexokinase is to attach a unit (called phosphate) to the fruit sugar, glucose, a favored food of the bacterium. We could compare this enzyme to a machine on an assembly line that attaches a bottle cap to a filled bottle of beer, but hexokinase is much more versatile. Although phosphate units are available within the bacterium, they cannot be connected directly to glucose. This step would go into the teeth of the Laws of Thermodynamics, which favor the removal of the phosphate. Instead, the phosphate is taken from a molecule named ATP (adenosine triphosphate), which has an internal energy supply. Transfer of a phosphate from ATP to glucose is favored as it is "paid for" by the energy present in ATP. But an ATP and a glucose wouldn't normally do business if they met on their own. The hexokinase acts as a broker, making the phosphate transfer comfortable for both participants.

In our animation, we would see some striking effects. A glucose (which we color green in our representation) might randomly bounce off the blue hexokinase until it hit the cleft. When it did so, the blue enzyme would suddenly change shape so as to grasp the glucose firmly in its "jaws." The glucose would remain trapped until a pink ATP wandered by. The enzyme would grab the ATP as well and place it near the glucose. The phosphate transfer from ATP to glucose would then take place, followed by another shape change by hexokinase. The products would then be released. The enzyme would return to its original shape, to wait for the next glucose. This scheme is clever in ways that are not obvious. The enzyme will not capture ATP unless glucose has first been bound. If ATP were bound alone, it might just drop off its phosphate, without any transfer to glucose. This would spend the chemical energy in ATP unproductively.

Hexokinase also contains a switch on its back that regulates the activity of the enzyme. The transfer of phosphate to glucose is the first in a long series of steps that ultimately converts glucose and oxygen to carbon dioxide and water. We could carry out the same change by setting the glucose afire, which would be about as useful as pouring gasoline on your car and igniting it. But just as burning gasoline in a fuel tank captures some of the energy and puts it to work, the slow burning of glucose in a cell does so as well. In a bacterium (or a human cell), the captured energy is then used to manufacture ATP. This molecule is used by the cell for energy transfers much as we use money. We spent one ATP in attaching a phosphate to glucose. But its ultimate combustion in a cell produces thirty ATP, returning the investment manyfold. When the cell has enough ATP, other uses can be found for glucose, and the regulatory switch is adjusted to lower the activity of the enzyme.

A chemist could not watch the operation of such a marvelous machine without wondering how it was built and why it worked. In our Museum, we can explore these questions with another animation. We would vibrate and shake the model to simulate the effect of heat on an enzyme. As we increased our energy input, the model would jiggle more and more until it suddenly went through a change—not a limited one like the opening and closing of the jaws but a more general unfolding process. At the end, the rigid compact structure would have converted to a loose rope that twisted and flopped about, keeping no particular shape. Needless to say, this rope would take no interest in glucose or ATP at all. When we turned off the vibrator, though, we would want the rope to coil back up and take on its original shape. This would mimic the remarkable effect that takes place with many real enzymes. When they are cooled slowly, they take back their initial form and regain their biological activity.

The loose rope form of our enzyme model will serve another purpose for us. When it is in its extended form, we can examine it more easily to understand how it is put together. On close inspection, we see that the strand is not uniform but resembles a charm bracelet made of interconnected links, with an ornament attached to each link. We have built it from hundreds of individual links, which can be connected or pulled apart. (In real life, a biochemist must use a treatment with hot acid or alkali to break the connections.)

When we have taken apart our model, we will be left with a collection of hundreds of separated links, each with at least two connection points and an attached ornament. Many of the links would be duplicates of one another, and if we sorted them out, we would find exactly twenty types. We would obtain a similar result if we had selected another protein from our bacterium or from any other living creature on Earth.

The connectors would come in two forms, somewhat like a plug and socket, with each link having at least one of each kind. The links could be plugged together again in any order whatsoever. In principle, we could plug them together again to form hexokinase, provided we had the original plan. If we lacked the plan, of course, we could not easily do so, no more than I could re-create a sonnet of Shakespeare that I had never seen before. We will stay with this point for a while, as it goes to the heart of the origin of life question.

The Connection Problem

> Humpty Dumpty sat on a wall
> Humpty Dumpty had a great fall
> All the King's horses and all the King's men
> Couldn't put Humpty together again.
> —*Nursery Rhyme*

Most proteins, or at least the smaller ones, can probably be constructed from scratch in a well-equipped laboratory today. The individual links, which are called amino acids, can be purchased or made fairly easily in the lab. The somewhat complicated procedures that must be used to snap them together have been perfected, and commercial machines called protein synthesizers have even been developed that can make the connections routinely. Apart from purchasing the supplies and servicing the machine, the biochemist need only type in the particular connection order that he or she wants, and collect the product. But this brings us to the heart of the problem.

All of the proteins made by living things on Earth are constructed out of a group of twenty related amino acids. In some cases, after the connection has been completed, some alterations may be made on the product. The finished proteins can then perform amazingly different tasks, within a bacterium or in the cells of our own bodies. Proteins drive muscle contraction, transmit nerve impulses, carry oxygen in our blood, support the structure of our nose, digest foreign materials, and help chemical reactions take place, in the style of hexokinase. Their shape controls their particular function, and the amino acid connection order governs their shape.

Our English language provides a good analogy. By using a word processor, we can type in characters and punctuation to form an extraordinary number of different meaningful messages, or just generate gibberish. With the same keyboard, we can reproduce the Ten Commandments, compose a love sonnet, or type out the weekly shopping and laundry lists. The order in which we strike the keys determines the result. If we hit them at random, we would be unlikely to get anything useful. This idea has been illustrated most often by imagining that a monkey was striking the keys. How long would it take the monkey to come up with the famous line from *Hamlet* (Act III, Scene 1): "To be, or not to be—that is the question"?

To do this calculation, we would have to answer a number of questions: Which keys would the monkey be allowed to strike? How far would he go before we recognized a failure and allowed him to start over? How fast can he type? It is easy to show that if we allowed him just one try, using 60 keys, his chances of getting it right would be incredibly unfavorable, about 1 in 1×10^{71}. To appreciate these odds, we can figure out how many attempts would be required for the monkey to have a reasonable chance to succeed. As a very rough rule of thumb, we will assume that the number of trials would need to be a number comparable to 1×10^{71}.

Recall that the number of stars in the galaxy approximated 10^{11} and that the number of galaxies in the universe was also of that magnitude. If we assume that every star in the universe has an Earth-like planet, and that the number of monkeys on each is equal to Earth's human population, and that all of the monkeys have been typing at one line per second since the Big Bang, then we end up with 2.3×10^{49} trials, a number that is still too small by a factor of more than 10^{21}. We would not expect to see our message unless we were very, very lucky. The point of this calculation is to show that the odds involved in generating short English messages can dwarf even those large numbers produced when we consider the dimensions of the universe. Similar considerations are involved when we think of producing hexokinase or another existing protein at random.

Astronomer Fred Hoyle and his colleague, N. C. Wickramasinghe, performed a similar calculation for the chances of combining the twenty amino acid subunits together to produce an enzyme with 200 amino acids in a particular sequence (connection order). Their calculation was simple and the results intimidating. The chances for that possibility were 1 chance in 1×10^{120}, far less than those for the sentence from *Hamlet*. Hoyle and Wickramasinghe then asked a broader question: What if we did not insist on getting the connections exactly right, but only wanted an enzyme that had the same activity, even if the sequence differed?

Biochemists have shown that proteins from different organisms can perform the same function, even if there have been a number of changes in the way the amino acids have been stuck together. In English, we can often get the same message across using a variety of expressions. With this in mind, Hoyle and Wickramasinghe reduced the odds to "only" 1 in 10^{20}. But their calculation applied to a chemist who had an instrument that linked amino acids together and also had preknowledge of the set of twenty that should be used. As we shall see, things are far worse if we were considering a random origin of life in a natural setting.

Despite these horrible odds, bacteria are well supplied with hexokinase and the other proteins that they need to maintain their existence. How do they get their supplies?

Bacteria Run Their Own Protein Synthesizers

When a bacterium needs a protein, it uses a strategy similar to one often used by the biochemist: it runs a protein-synthesizing machine. The microbe, however, stores the apparatus safely in multiple copies within its own body.

The technical name for this machine is the *ribosome*, and a typical *Escherichia coli* cell may contain 20,000 of them, which make up a substantial proportion of the cell's weight. We will reserve a separate room in the imaginary Museum of the Cosmos, at the same -8 level, for a ribosome model. As we entered we would confront an irregularly shaped, roughly spherical object with its diameter roughly that of the height of a basketball player. When we looked more closely, we would see that it was made of two large subunits that resembled a baseball mitt holding a "rubber duckie."

Each of the two parts would present itself as a quilt of blue and pink areas. As we have learned, blue represents protein. We have selected pink for another cell construction material: nucleic acids. To learn more, we will push a button that animates our model and watch the two parts separate, and then each part come apart into a host of components. At the end, we would have a jumble of fragments, as if we had disassembled a three-dimensional puzzle. A quick count would reveal that we had over fifty different proteins and three different nucleic acids. Each of the proteins would itself be made of connected amino acid links, as they were for hexokinase. The nucleic acids would make up by their bulk what they lack in numbers. Much more pink than blue would appear in our model because two of the three nucleic acid components are huge, when compared to the proteins.

Our model has been constructed quite cleverly and can perform more tricks. By pushing another button, we can switch on the vibrating and shaking movements that mimic the effect of heat. Each of the host of proteins will unfold to a loose rope, as we saw for the case of hexokinase. Furthermore, the nucleic acids will do so as well. If we inspected one of them more closely, we would see that it is made out of connected links joined in sequence, just like a protein. What would differ is the chemical nature of the links, which are called nucleotides. Only four are used, rather than twenty, as in the case of protein, and each nucleotide is much more intricate in its construction than an amino acid.

Another difference between the proteins and the nucleic acids is that the latter come in two different styles; think of two-door and four-door sedans or caffeinated and decaffeinated coffee. Our ribosome has been equipped with RNA (for ribose nucleic acid). We shall encounter the other selection shortly.

If we despair because our intricate ribosome model now lies as a jumble of parts on the floor, each in its unfolded form, we need only push the buttons once again. As if by magic, the pieces fold up, locate one another, and reassemble to form the original structure. Something not too different actually takes place within a bacterium, and the process can be demonstrated in the laboratory.

If we were uninformed, we might speculate that each ribosome specializes in the manufacture of a particular protein, but in practice they all operate as all-purpose machines, just as the protein synthesizer does in the lab. In the latter case, the scientist looks up the desired connection order in his or her notes and feeds the information into the machine. In the case of the ribosome, a tape carrying the necessary information threads through it. As the tape is made of RNA, a subclass that carries the descriptive name messenger RNA, we will color it pink. This RNA has been connected by the same plan that was used to make the RNA within the ribosome, but the connection order of the subunits is different.

The messenger RNA sequence determines the protein that is made by a particular ribosome, much as the tape or CD that you put into an audio system determines the music that comes out. In each case, the machine must read the information that has been fed in and alter it to produce output in a different medium or language. The language used to store information in a nucleic acid differs from the one employed in protein. The ribosome reads the messenger tape, which is written in the language of RNA, and interprets it in terms of a protein sequence. Naturally, geneticists have named this process "translation."

We have penetrated one of the deep secrets of the living process, but another question comes up immediately: Where does the messenger get its message?

The Source of the Information

We will return to the biochemist who was preparing a protein in a laboratory. How did the biochemist obtain the connection order that he or she typed into the synthesizing machine? If the protein was a familiar one, then the biochemist probably found its desired sequence in the library. In the same way, the messenger RNA of the bacterium brought the information to the ribosome from a central storehouse, one made of DNA, the other type of nucleic acid called DNA.

To visualize this biological information depot, we will return to the main display area at the −8 level, where we keep the ocean-liner-sized model of an entire bacterium. We will switch on a sound and light system that darkens most of the display and illuminates one special structure. An enormous pink coiled rope about 20 centimeters thick occupies much of the inside of our bacterial model. This structure, the chromosome, functions as the central library and holds the information needed to make all of the proteins used by the bacterium. A messenger takes only a limited amount of information from this library, which stores it all on one long tape.

In 1998, after an effort of almost a decade, the total information content of the *Escherichia coli* chromosome was read out for the first time. In that body, 4,639,221 letters of DNA language were encoded, with the information for 4,288 different proteins. The book that you are reading now holds only about one-fifth of that number of characters. Had Henry Bastian realized this, he might have discontinued his experiments on spontaneous generation with hay infusions.

A bacterium functions by drawing on this internal treasure of stored information, which represents a huge accumulation of negative entropy. But where did it get this stockpile?

As Pasteur demonstrated, the organism did not create it anew, but inherited it from earlier life. Bacteria reproduce by cell division—splitting in two. When that event takes place, the chromosome is duplicated, and each daughter cell receives a copy of the DNA sequence, the heritage of its species. When we seek the source of the information, we find that the trail leads back into the indefinite past, affording no clue as to how life began. Yet more than a century after the experiments of Pasteur, we find scientists again attempting to demonstrate a start through spontaneous generation.

5

The Missing Machine

Scientists avoid the supernatural in seeking to explain the origin of life. In the nineteenth century, they tried to show that bacteria could be formed in a few days when a sterilized solution of yeast extract or hay infusion was left to sit undisturbed. They worked in vain. We now understand that such an event was very unlikely, because the simplest living organisms are still incredibly complex. Physicist Harold Morowitz has provided calculations that show why the most confirmed lottery addict would refuse such a wager.

The Odds of Morowitz

Professor Morowitz has calculated the chances for success in the following experiment. Suppose that a single bacterium was heated up several thousand degrees, so that all of its chemical bonds were broken. The resulting mixture of hot atomic gas was cooled slowly, to allow bonds to form once again by random processes. What would the odds be that the chemicals would recombine in a way that restored the living bacterium? The number calculated by Morowitz was $10^{100,000,000,000}$ to 1. That number represents a 1 followed by 100 billion zeroes. If we simply wanted to write that number without using exponents, we would need to use several hundred thousand volumes. The first book would start with a 1, with zeroes filling that page and all of the remainder of them. The other volumes would contain only zeroes. I doubt that even Jacques Monod would have wanted to assume that life originated in this manner, in the face of such odds. We must assume that life began with a much more favorable process.

At present, proteins and nucleic acids (RNA and DNA) cooperate hand in hand in controlling the central processes of life. Other chemicals help as well, such as the lipids that make up much of the bacterial cell's covering. Could life function with just one of them? Many scientists have convinced themselves that the answer must lie in this direction. They have been inspired by a series of landmark developments in modern science, which started with Charles Darwin's theory of evolution.

In the Beginning: Natural Selection

The fossil record reveals that a succession of creatures have inhabited the Earth during its history. Their geological position and radioactive dating allow us to arrange their appearance in order, with rough dates. But this evidence does not tell us how new species were generated from earlier ones. Darwin's theory emphasized that a single mechanism was responsible: natural selection. We have been plagued by an unwanted example of this process in recent years, in the appearance of bacterial strains that resist antibiotics.

When penicillin was first introduced, it devastated the microbial populations, which were not prepared to defend themselves against it. This drug interfered with their ability to construct cell walls, with lethal effects. But one day a novel microbe appeared that resisted penicillin's action. That favored individual had gained the ability to produce a protein that destroyed penicillin. As a result, it alone survived and multiplied. Its descendants now contained a new gene that afforded them resistance to this drug (a gene is the term used for a stretch of DNA that holds the instructions for a particular protein).

That new protein did not result from a frenzied round of bacterial research but appeared accidentally. When the master tape of DNA is copied during reproduction, some errors are introduced at random, just as we introduce mistakes when we copy a manuscript. No process is perfect. If the change occurs in the part of DNA that describes a vital protein, and it takes place in a way as to render the protein product useless, that individual dies. Such a change is called a lethal mutation. Occasionally, a favorable mutation occurs, one that gives an existing protein new and desirable properties.

No bacterial holiday has been declared to mark the day and record the events, but we presume that such events took place when penicillin resistance first arose.

I have described a process in which new genes arose through mutation. Darwin wrote in terms of the development of new species, which presumably involved multiple genetic changes, but species development was still governed by the same process of natural selection. Gregor Mendel, Thomas Hunt Morgan, and many other dedicated workers had to labor over decades to uncover the mechanism of genetic change. Finally, in 1953, James Watson and Francis Crick proposed a three-dimensional model for DNA that provided insights into the way DNA could store information and be copied. A single molecule lay at the heart of heredity. Many scientists found it irresistible after this discovery to place the gene at the center of the life process, perhaps capable of life on its own.

For example, Hermann J. Muller, the Nobel Prize-winning geneticist who discovered the effect of radiation in causing mutations, wrote in 1966, "The 'stripped down' definition of a living thing offered here may be paraphrased: *that which possesses the potentiality of evolving by natural selection. . . .* The gene material also, of natural materials, possesses these faculties and it is therefore legitimate to call it living material, the present day representative of the first life."

Astronomer Carl Sagan, as a graduate student, had earlier speculated on the possibility of "a primitive free-living naked gene situated in a dilute medium of organic matter." Evolution then represented the extension of the gene's ability to provide for its future. As Richard Dawkins argued in *The Selfish Gene,* the bodies of animals are "survival machines" for the genes within them. The body of an elephant is just an elaborate device that ensures the perpetuation of elephant DNA.

Evolution in the Test Tube

These speculations were given a boost by an elegant series of experiments performed by biochemist Sol Spiegelman and his colleagues in the late 1960s and early 1970s. They attempted to show that a "naked gene" could adapt to changes in the environment through suitable mutations. They selected their gene from a virus, $Q\beta$ by name, that normally dwells within bacteria and functions by borrowing most of the necessary cellular machinery of its host.

The genetic material of this virus is relatively small, constructed from RNA rather than the DNA used by all cells, and contains the information for only a few of its own genes. One of the genes specifies a protein called Qβ replicase, which is essential because it is needed for the RNA gene to be copied.

Spiegelman created a novel and minimal world for his chosen naked genes. He gave them an abundant supply of the nucleotide subunits that they needed to build copies of their RNA; these building blocks had a built-in energy source that favored their connection. He also provided the replicase. The RNA now had everything it needed to be fruitful and multiply, and it did so, for generation after generation. What could limit this process of endless reproduction? Eventually, the subunit supply would be exhausted, or the container would run out of space. The researchers worked around this problem by transferring a sample of the mixture, every so often, to a new flask with fresh subunits and replicase.

During this process, errors crept in and mutant RNAs were produced. These new "species" competed with each other in their reproduction ability. If one could be copied in a shorter time than its siblings, it could have a greater number of generations within a given time and would grow in proportion in the mixture. For example, if a mutant RNA could be copied in ten minutes rather than the usual twenty, it would have two generations and four descendants in twenty minutes. The original RNA would have only two offspring after that time and would fall farther and farther behind with the passage of additional time.

The most dramatic way for the RNA of virus Qβ to speed up its copying time was to shorten its length. A short message can be typed more quickly than a long one on a word processor, and the same rule applies to the copying of RNA. The information that the RNA carried was relevant to its life cycle within bacteria, not to its present circumstances. Only one property was needed for reproduction in Spiegelman's world: RNA had to be recognized by the replicase. That enzyme had a property essential for the normal life cycle of the virus. It copied Qβ RNA but ignored the many RNAs produced by the host bacterium. If it could not do this, then almost all of its time would be spent in needlessly copying the abundant host RNA, and the virus would never get to reproduce.

In the test tube, random breakage events would shorten the Qβ RNA. For a fragment to remain in evolutionary competition, it had to retain intact that portion of its message that qualified it for reproduction by the replicase. In its experiment, the Spiegelman group followed the process of change of the RNA for over seventy generations. At the end of that period, a single descendant dominated the mixture. It had started with about 4,500 nucleotides; now only 550 remained. Most of the original RNA had been discarded, to speed up the copying process.

Another set of experiments was run using $Q\beta$ RNA that had already been shortened. A drug was added that bound to certain sites on the RNA, greatly slowing down the speed of copying. A number of generations were followed, in the presence of the drug, keeping the same experimental procedure. At the end of that time, a mutated champion again dominated the mixture. The winner had undergone a change in a single "letter" of its RNA sequence at three places. These changes destroyed the favorite binding site of the drug that they had added. As a result, the reproduction rate of the mutant had returned almost to the one that existed when no drug was present.

Many experiments with this system and related ones were carried out afterward by Nobel Laureate Manfred Eigen and others, to understand the selection process in detail. Such experiments have been called "evolution in the test tube," as they demonstrate a simplified molecular selection. But the setup still relied on an important ingredient that is absent in nature: the human experimenter.

The experimenter had to supply the replicase and building blocks that the RNA required for its proliferation; the RNA did not make its own way in the world. The system represented an elegant model but could not tell us what happened early in evolution. To extend these ideas to the origin of life, scientists had to confront an ancient question: "Which came first: the chicken or the egg?"

RNA (or DNA) was the "egg." It could carry information and be copied but could not replace proteins in the everyday functions necessary to life. The RNA in Spiegelman's system could not even reproduce without help from the replicase, which was a protein. Protein molecules (the "chicken") could do the work, but they were not constructed in a way that allowed for them to reproduce. They were sterile. The puzzle appeared insoluble unless one of the premises was wrong. In 1968, Francis Crick and Leslie Orgel suggested that RNA could do some of the work after all and operate without protein. This idea rested in limbo until some new experimental results came in.

The Birth of *"RNA World"*

For those who endorsed this reasoning, a critical breakthrough took place in the 1980s. Colorado State chemist Thomas R. Cech found that RNA, under some circumstances, could act in the manner of a protein. For this insight, he and Yale biochemist Sidney Altman shared the 1989 Nobel Prize.

Their discovery was accidental—certain natural RNA molecules were found to have the property of cutting and splicing themselves without the help of a protein assistance. The splicing process was part of their normal maturation within living cells. The name "ribozyme" was coined to describe the newly discovered class of RNA enzymes.

Heroic efforts were undertaken to find other things that RNA molecules could do, apart from cutting themselves. Some workers sought RNAs that could function as medicinal agents, by destroying viruses. But for others, the discovery of a ribozyme that could act as a replicase became a prominent goal. If such a replicase were available, Spiegelman-Eigen-type experiments on evolution in the test tube could be conducted in a much more interesting way. An RNA molecule would be copied while an RNA relative (or identical twin) acted as the midwife for reproduction. Mutations would affect both functions, and novel and unexpected results might emerge, shedding new light on natural selection. A number of scientists felt that such experiments would provide the key clue to the origin of life.

A germinal paper was published in 1986 by Walter Gilbert. He had shared the Nobel Prize in chemistry with Fred Sanger for the development of methods to read out the information stored in the DNA in living organisms. With early training in physics, Gilbert had an inclination to theoretical speculation not usually found in biologists. In his paper he combined the earlier naked gene idea with new information about ribozymes and applied them to the origin of life: "The first stage of evolution proceeds, then, by RNA molecules performing the catalytic activities necessary to assemble themselves out of a nucleotide soup."

He applied the name "RNA world" to his vision of a biosphere in which RNA performed all the key functions before proteins entered the scene. The name and the idea caught on. For an example, I need only turn to the very widely used textbook from which I teach biochemistry. It features a section with the heading (in capital letters): "RNA probably came before DNA and proteins in evolution." Within the text, the author writes of an ancient epic "that probably began when RNA alone wrote the script, directed the action, and played all the key parts." Many other sources have reacted in the same way, some dropping the word "probably."

Efforts to fill those key parts, including the replicase, have moved slowly, however. Nature has not been helpful, leaving no trace of most of these presumed players. In the absence of any clue, scientists have attempted to prepare an RNA replicase on their own and have run head-on into the connection problem. The number of possibilities to be examined in this hunt is staggering. For example, if scientists wished to prepare a mixture that contained one molecule of every possible RNA with a length of 100 nucleotides, at the concentrations usually chosen for biochemical research, they would need a container about eight times the diameter of our solar system to hold it. Obviously, some shortcuts are needed. Biochemists such as Jack W. Szostak at Massachusetts General Hospital and Gerald F. Joyce at the Scripps Research Institute in La Jolla, California, have been ingenious in discovering them.

Using commercial synthesizers, protein enzymes, and a host of elegant techniques, researchers prepare more limited RNA mixtures with "only" 100 trillion molecules. Elaborate multistep selection techniques are used to separate molecules that may have some abilities of the type they are seeking. These candidates are then allowed to multiply, using protein replicases. Their connection order is deciphered, and the information is used as the starting point for the next wave of experiments.

In this quest for new kinds of ribozymes, good progress has been made in isolating species that resemble ones already in the menagerie. Gerald Joyce, for example, has prepared a ribozyme that cuts up DNA rather than RNA. The search for an RNA replicase has gone more slowly. I will let Dr. Joyce present the adventure in his own words:

> If one believes that an RNA-based life form is possible, then why not make one in the laboratory? . . .
>
> A research biochemist knows how to obtain the components of RNA. They can be bought from a chemical supply house! These components are available as pure compounds having only the proper handedness. They can be assembled in the laboratory to produce RNA. The challenge is to devise RNA molecules that have the ability to direct their own replication. . . . RNA evolution can be made to occur, leading to the evolution of new and interesting RNAs whose functional properties conform to the demands of the experimenter. . . .

It is probably only a matter of time, to be measured in years rather than decades, before a self-sustained RNA evolving system can be demonstrated in the laboratory. This would be a case in which a DNA- and protein-based life form, namely a human biochemist, gives rise to an RNA-based life form, an interesting reversal of the sequence of events that occurred during the early history of life on Earth.

When that event takes place, the media will probably announce it as the demonstration of a crucial step in the origin of life. I would agree, with one modification. The concept that the scientists are illustrating is one of intelligent design. No better term can be applied to a quest in which chemists are attempting to prepare a living system in the laboratory, using all the ingenuity and technical resources at their disposal. Whether they use synthetic chemicals or materials isolated from nature, we would be justified in calling the living system artificial or human-made life.

Literature preceded science in this area, with Mary Shelley's 1818 conception of Dr. Frankenstein's monster. But the novel *Frankenstein* led us emotionally in the wrong direction. The first artificial living system would, by necessity, be far simpler than a bacterium, and I doubt that we would have much to fear from it. It would reflect a triumph in our understanding of nature and a testimony to our technological ingenuity. But it would tell us essentially nothing about how life first started on this planet. There were no biochemists or modern laboratories on the early Earth.

The search for ribozymes evokes the same feeling of achievement and beauty in me that I get when I see a skilled golfer playing a difficult course at well under par. To imagine that related events could take place on their own appears as likely as the idea that the golf ball could play its own way around the course without the golfer. We can, of course, imagine that natural forces would lend a helping hand. A hurricane could move the ball down the course, and occasional floods might "putt" the ball into the hole. A small earthquake could then remove it and place it on the next tee. Perhaps each of these events could be simulated if we tried hard enough. But to insist that all of these events be linked together and move in an appropriate direction puts our origin into the realm of Morowitz's odds. We would be left in the Sour Lemon School position: The origin of life was an unlikely, once-in-a-universe proposition.

If we want to model how life may have started on the early Earth, we must listen to nature instead of instructing it. By appropriate simulations, we might gather some clues about events at the beginning. In fact, an entire discipline called "prebiotic chemistry" has attempted such investigations for much of the past century. Such investigations have had ample media coverage, and the bulletins that emerged have often given the impression that the origin of life problem is all but solved. As we shall see, I disagree with this analysis. But first, we should present the position.

I have invented a spokesman for this purpose who I will dub (with apologies to the famous author of children's books) Dr. Soup.

An Interview with Dr. Soup

Soup's the name; Primordial Soup if you want my full name, but you can call me Primo. Sometimes I use the name Prebiotic Soup, but if it gets too confusing, just call me P. Soup.

I was "born" in the old Soviet Union, just after their revolution, in 1924. My "father," so to speak, was a biologist, Alexander Oparin, who is also called the father of the origin of life field. My birth announcement was published in a Moscow botanical journal and, as you can imagine, didn't draw much attention. A famous English population biologist, J. B. S. Haldane, came up with the same idea on his own, but Oparin stayed with it and publicized it, so he got the credit. By the late 1930s, his ideas had expanded into a book, which was translated into English. His theory attracted more attention then.

I could explain it to you on my own, but the writers have done such a good job that I couldn't improve on it. Let me read from one of my favorite versions, the one given by astronomer Robert Jastrow in his best seller *Until the Sun Dies:*

> The earth is one billion years old. The sky seems familiar; its color is a deep blue, spotted by puffs of white cloud. But its gases are strange; in place of oxygen, the atmosphere contains pungent fumes of ammonia, the odorless menace of methane, and traces of hydrogen. A shallow sea covers the planet. Its waters are sterile; life will flourish in them later but has not yet appeared. . . .
>
> Now a thunderstorm lashes the surface of the planet. . . . In each electrical discharge, the gases of the atmosphere—methane, ammonia, water, hydrogen—fuse together to form strange new com-

binations of atoms, not previously seen on the earth. These groups of atoms are the molecules known as amino acids and nucleotides.

Gradually, the amino acids and nucleotides drain out of the atmosphere into the oceans, creating a rich soup of organic matter, like a chicken broth but more concentrated [and] . . . during the course of a billion years, every conceivable size and shape of molecule is created by random collisions.

Eventually, after countless millions of chance encounters, a molecule is formed that has the magical ability to produce copies of itself. . . . The original molecule was the parent; the copies are its daughters. In a short time they dominate the population of molecules in the waters of the young earth. . . . When the first DNA-like molecule appeared, the threshold was crossed from the non-living to the living worlds. . . . During the billions of years that followed, these simple self-reproducing molecules evolved into the variety of plants and animals that now populate the earth.

Dr. Soup put the book down and continued on his own:

Of course, Oparin's original theory didn't tell about nucleotides and DNA. He had to work within the Soviet theory dialectical materialism, which didn't believe in the importance of genes. The genetic part was added by scientists in the West, and Haldane accepted that concept.

The theory really came into its own when a famous experiment was reported in 1953. The Nobel Prize-winning chemist, Harold Urey, wanted to test the ideas, so he recruited a bright young graduate student, Stanley Miller, to build some apparatus. Both of them were at the University of Chicago at that time. Miller ran an electric discharge through the mixture of gases that were said to be present when the Earth was young and trapped the products in water. What do you think he got? Amino acids, just like the theory proposed.

"*Time* magazine picked up the story," continued Dr. Soup, who then read from an old 1953 issue:

[Miller and Urey] had simulated conditions on a primitive Earth and created out of its atmospheric gases several organic compounds that are close to proteins. . . . What they had done is to prove that complex organic compounds found in living matter can be formed. . . . If their apparatus had been as big as the ocean, and if it had worked for a million years instead of one week, it might have created something like the first living molecule.

Other papers and magazines carried the news as well, Dr. Soup explained. "By now, you can find this story everywhere: museums, textbooks, and a host of technical papers."

I had to interrupt Dr. Soup at this point for a question. I wondered why the primordial soup needed to occupy an ocean, and why millions of years were required. He responded immediately. "That's because you have to connect the smaller pieces together properly to form the first living gene, which requires a bit of luck," Dr. Soup countered. "That's why you can't show it in the laboratory today. No graduate student is willing to spend a million years on an experiment. Professor George Wald, another Nobel Prize winner, caught the idea perfectly." Dr. Soup picked up an old *Scientific American* and began to read from Wald's article:

One has only to contemplate the magnitude of this task to concede that the spontaneous generation of a living organism is impossible. Yet we are here—as a result, I believe, of spontaneous generation. . . . Time is in fact the hero of the plot. The time with which we have to deal is of the order of two billion years. What we regard as impossible on the basis of human experience is meaningless here. Given so much time, the "impossible" becomes possible, the possible probable, and the probable virtually certain. One has only to wait: time itself performs the miracles.

George Wald really could turn a fine phrase, Dr. Soup added with envy, but he wasn't the only one. Soup picked up a geology text and continued:

How many times 10,000 trials of such random events could have occurred within a period of 3.3 billion years? One's imagination boggles at having to calculate so great a number. No one familiar with statistics rejects the idea of chance chemical combinations because there wasn't enough time. There was a huge abundance of it.

Another question troubled me. Miller had obtained some amino acids in his experiment. They are the subunits of proteins. For many years, biochemists had thought that proteins were the materials of heredity. But in

1944, Walter Avery and his colleagues found that genes were made of DNA. Within a few weeks of the date that Miller published his paper, James Watson and Francis Crick published their impressive structure for DNA, which showed how this chemical could function as a gene. But one needed nucleotides, not amino acids, to construct DNA or RNA. I knew that nucleotides were far more intricate than the amino acids that Miller had produced.

To form a nucleotide, you have to connect three different chemicals in a very specific way: If you wanted a building block for RNA, for example, you had to select one of four information units (bases): adenine, cytosine, guanine, or uracil, and attach it in an unusual way to the sugar, ribose. The product then had to be united with the mineral, phosphate, to form a nucleotide. In spite of these difficulties, RNA world advocates assumed that the early Earth provided an abundant supply of such substances. To cite one source out of many, Manfred Eigen and his colleague Peter Schuster wrote in 1982, "The building blocks of polynucleotides—the four bases, ribose and phosphate—were available too under prebiotic conditions. Material was available from steadily refilling pools for the formation of polymers, among them polypeptides and polynucleotides." (If some of the words are unfamiliar to you, substitute *RNA* for *polynucleotide* and *protein* for *polypeptide.*)

I asked Dr. Soup what the basis was for such optimism. Had Miller's experiments produced a bumper harvest of nucleotides in his simulated ocean? Had these substances been found in nonliving sources such as meteorites? He was quick to respond.

> No, they haven't been found there at all. But many chemists have shown that it's not a problem. They can make these compounds in their labs under "prebiotic conditions," which mimic those of the early Earth. Juan Orö has shown that you can make adenine and guanine by heating concentrated cyanide solutions, for example. We've known that you can make ribose just by heating up some formaldehyde. Cyanide and formaldehyde are simple chemicals; surely there was a lot of them around. And phosphate is present in many rocks. Cytosine and uracil seemed a problem for a time, but Stanley Miller solved that just a while ago. We still haven't worked how to connect them to the sugar, but some bright fellow will come up with a clever idea sooner or later.

Once we have made the nucleotides, we then have to connect them together somehow. Jim Ferris has done a first class job in this area. He's at Rensselaer Polytechnic; he's won the Oparin Medal for his work. Ferris was able to grow pieces of RNA that are over fifty units long, working with mineral surfaces. So we just need to find the right formula for the replicase. We might need to connect as few as thirty units together to get that activity. The total number of combinations of that length is a big number, but Joyce and Szostak can handle it. So we have every reason for optimism. The answer is just around the corner.

Dr. Soup looked very content as I ended the interview.

A Skeptic's Complaint

At this point my narrative gets "up close and personal" as we enter an area that I know extremely well. In composing a presentation for Dr. Soup, I tried to blend the ideas put forth in papers and at meetings by workers in the field of prebiotic synthesis, particularly those who believe that life began with an RNA world. I will quote Jim Ferris for the underlying philosophy: "The first life forms probably contained nucleic acids for the storage of genetic information. Consequently, there must have been a pathway by which the nucleotide building blocks of RNA were synthesized."

These scientists have tried to identify candidate chemical reactions that lead to RNA under conditions that may have existed on the early Earth. They run their reactions in water, avoid strong acids and alkalis, and use chemicals that they consider "prebiotic" for their experiments. To qualify as prebiotic, a chemical must appear in a Miller-Urey-type reaction or be produced in another "prebiotic" experiment. The scientists have assembled lists of such reactions, which they feel supports their general position. They admit that problems exist but feel that they can be solved by additional work on their part. As Jim Ferris stated, the pathway *must* have existed. The scientists' job is to locate the correct one.

My own opinion has been very different: These reactions, while well carried out in most cases and often ingenious, have nothing whatsoever to do with the origin of life. I have brought this message to such scientists at conferences, in my book *Origins: A Skeptic's Guide to the Creation of Life on*

Earth, and in technical papers published in the official journal of the International Society for the Study of the Origin of Life. In return, my criticisms have been called a "complaint," and I have earned the unofficial title "Dr. No" from one of the prominent laboratories. My views, though, come not from last night's poorly digested dinner but from a lifetime of work with DNA and RNA. From my own work and the reading of thousands of articles, I know that DNA and RNA are very hard to put together, but easy to take apart. Their formation on the early Earth would require a set of unlikely events that would put our origin in tune with the Sour Lemon philosophy. One other aspect of my scientific training has also led me to my particular perspective.

I obtained my doctoral degree at Harvard under a great master of organic synthesis, Robert B. Woodward. He earned the Nobel Prize in chemistry by demonstrating that he could prepare some of the most intricate products of nature in his laboratory. His conquests included quinine, strychnine, lysergic acid, chlorophyll, and vitamin B_{12}, which he carried out in collaboration with a brilliant Swiss organic chemist, Albert Eschenmoser.

In preparing a natural substance for the first time, Woodward and other chemists attempted to carry out a "total synthesis," which meant that they could only use chemicals that had also been prepared in the laboratory. This limitation carried an advantage with it, however. Once a chemical had been prepared in that manner, even in the tiniest amounts, it could then be obtained in unlimited quantities from any source whatever for further work.

An example will make this clear. Let us suppose that Alice and Bill are the names of chemical intermediates that are needed for a total synthesis of chlorophyll. Another chemist worked out a synthesis for Alice ten years ago, and it is now produced in a factory. You are permitted to order a ton of Alice from the factory and use it to start your own synthesis. Bill, however, has been isolated only as a breakdown product of chlorophyll. Since chlorophyll has not yet been synthesized, you cannot use Bill at this point.

But assume that after a year of work, you have converted your ton of Alice to a trace of Bill, enough to identify it but not enough to do anything with it. Just the same, you have made a huge advance. Bill has now been synthesized, so you are free to get it as best you can. Chlorophyll is abundant in nature, as the green pigment of plants. You are allowed to degrade all of the natural chlorophyll you choose to produce your supply of Bill. When a compound such as Bill is used in this manner, it is called a *relay*. If you can then reconvert the smallest fraction of that Bill to chlorophyll, you can claim a total synthesis of chlorophyll.

I have taken this side excursion for a reason. Ideally, prebiotic chemists should imitate the conditions of the early Earth as best they can in their laboratory, then step aside and observe what happens. What they have done instead is attempt a total synthesis of RNA, in the style I have just described. One early target for synthesis, for example, is adenosine, a combination of the sugar ribose with the base adenine.

As Dr. Soup mentioned, ribose can be prepared by heating formaldehyde, usually in alkali. Unfortunately, the mixture produced is a total mess. You get some ribose, but in small amounts along with scores of closely related substances. The same is true for adenine production from solutions of cyanide; hordes of products are produced. Prebiotic chemists, however, then take pure adenine and heat it with pure ribose under a new set of conditions. The reaction goes poorly, but some adenosine may be produced. This result qualifies as a prebiotic synthesis of adenosine. It is now legitimate to use high concentrations of pure adenosine for the next step. Adenine, ribose, and adenosine have been used as a relay, a tactic that is permissible in total synthesis but ludicrous as a model for our early planet.

It would be much more realistic to heat together the entire formaldehyde and cyanide products, which would furnish the mother of all messes. Better yet, the chemist should simply mix the cyanide and formaldehyde starting materials. But we know what happens in that case; the two substances have a great affinity for each other and their reaction takes off in a direction that bears no resemblance to life as we know it.

This example does not represent an isolated lapse. If one digs into the literature behind almost any of the prebiotic claims that buttress RNA world, one finds no greater degree of merit. We have much ground to cover elsewhere, and the planets clamor for our attention, but I cannot resist bringing up a single additional case.

Miller's Lagoon

The news reached me first at home as I read my *New York Times* on Independence Day, 1996. The headline proclaimed, "Chemist Adds Missing Pieces to Theory on Life's Origins," while the subhead added, "All four chemical bases of RNA can arise in nature." A box associated with the article echoed the message: "It is now known that all four components of RNA can be produced by natural processes on the face of the earth, a finding that has profound implications for scientists' thinking about the origins of life." An accompanying photograph showed an unhappy-looking Stanley Miller standing before his apparatus. He had continued in the field for over thirty years since his celebrated amino acid experiment, and he had a laboratory at the University of California at La Jolla, near San Diego.

My attention had certainly been caught. I had just published a review about the "prebiotic" claims for adenine, one of the four RNA bases. I had come to the conclusion that only traces of it, at best, would be found on our planet before life began. My review had been published in *Origins of Life and Evolution of the Biosphere,* the official origin of life journal, with hardly any change requested by the referees and by the editor, Jim Ferris. Stanley Miller had invited me to present the information to the American Chemical Society in San Diego, which I had done. Still, two years had passed, and there was plentiful room and pressing need for some new discoveries.

There was no new information about adenine, however. The *New York Times* account announced a new "prebiotic" synthesis for two other RNA bases, cytosine and uracil:

> Both substances might have been produced by the lifeless young oceans in ample quantities by a process involving the evaporation of sea water in tropical lagoons, the freezing of sea water in polar regions and the mixing of their products in the open ocean. . . . The evaporative part of the process, Dr. Miller said, could have concentrated the traces of urea that accumulate in sea water as a result of reactions in the atmosphere caused by lightning flashes.

As the details were somewhat skimpy, I waited for the latest issue of the prestigious journal *Nature* to arrive; it carried a more detailed report. Only the evaporating lagoon scenario had an experimental basis; the other was simply mentioned in passing. Miller and his co-worker, Michael P. Robertson, had written that "here we show that in concentrated urea solutions—such as might have been found in an evaporating lagoon or in pools on drying beaches, cyanoacetaldehyde reacts to form cytosine in yields of 30–50%, from which uracil can be formed by hydrolysis."

Nature had given its blessing to the claim by reprinting it on a page devoted to the highlights of each issue and adding the summary, "These reactions provide a viable route to the raw materials required in the RNA world."

Had a vexing problem in the origin of life been solved so readily? The prospect provoked me into a reading program on lagoons, about which I knew little, to e-mail correspondence with several planetary geologists, and to further investigation on the chemistry, which was more familiar to me. My conclusions, which were highly pessimistic, have been sent to an appropriate journal, but they are written in technical language. Here I would like to try an alternative presentation, one that starts with a positive frame of mind. I have borrowed my presentation style from the Fundamentalists.

The Creationists presume that the Bible is literally correct and construct a view of the universe that harmonizes with that assumption. If their view requires that humans and dinosaurs coexisted, and that the speed of light was once infinitely fast, then so be it. In this spirit, let's again declare that there must have been a way that RNA arose on the Earth, before life began. Since cytosine is one of the essential components of RNA, it must have been present in ample quantities, to encourage RNA construction. Only the Miller-Robertson preparation offers high yields, so we will assume it to be the one that actually operated on the early Earth.

Miller and Robertson did not experiment with lagoon simulations, however. They ran their reactions in the laboratory using pure, concentrated chemicals and sought the highest possible yield by varying the conditions. I will try to transpose their results to the early Earth and deduce what conditions would be necessary there to make the experiments work.

To start, I questioned first why lagoons were required. Most descriptions of the prebiotic soup have allowed it to cover the globe. Why not assume that the reaction took place in open ocean?

Unfortunately, this would not work. Miller himself and and other scientists have calculated that concentrations of most chemicals in the ocean would be very low. Urea is prominent in our internal body chemistry, and in our urine today, but it is produced only in modest amounts by lightning spark discharges. We would not expect much of it to accumulate on a global basis. For the new reaction to succeed, we need very high concentrations of urea. To get them, we would need to evaporate an amount of ocean water sufficient to fill a pond down to the size of a bathtub, concentrating it to one-millionth of its initial volume. The concept of drying lagoons comes immediately to mind.

In my reading I learned that lagoons are plentiful on Earth today. Some that are surrounded by coral reefs are irrelevant, as they are products of life. Many others are formed, however, when bodies of water are partly cut off from the sea by sand bars or other natural obstructions. If I wanted

to drive perhaps half an hour from my home to the south shore of Long Island, I could inspect a large lagoon to my heart's content. But my principal
source book on lagoons also warned me that "lagoons are rarely completely isolated from the sea. Characteristically, they have a channel (or
series of channels) through which water is exchanged with the larger adjacent water body."

In those cases where a lagoon was completely cut off, it usually evolved
into a freshwater pond or lake. Alternatively, the ocean might break the
barrier and reestablish the connection. The geologists whom I corresponded
with couldn't furnish an example of an isolated lagoon that had evaporated
almost to dryness.

On the positive side, less complete evaporation had been documented.
This took place in arid areas of the globe between latitudes 15° and 35°.
From 2 to 8 meters of liquid could be lost per year under such conditions.
A series of salts would then be deposited as the volume contracted to as
little as 5 percent of its initial amount. One advantage of urea is its great
solubility, which would protect it from this fate. But this benefit was counterbalanced by a disadvantage. Urea is unstable over long time periods,
forming ammonia and carbon dioxide. Stanley Miller and Leslie Orgel have
estimated that half of the urea decomposes in twenty-five years at 25°C
(77°F). If the evaporation of the lagoon went slowly, and was occasionally
reversed by rain and the incursion of rivers, all would be lost. The synthesis would require a very broad, shallow lagoon, in a very arid and rainless
climate, with continuous winds to speed the evaporation.

If partial evaporation could take place under such circumstances, why
didn't it go further to provide the extreme concentration needed for the reaction? One reason was that the rate of fluid loss slowed as the liquid grew
more thick. But also, as the liquid grew more dense, its level simply sank
below the sandy floor of the lagoon, protecting it from further evaporation.
We have to specify further that the lagoon have a rocky, impermeable floor
rather than a porous one. There are no data on the occurrence of such lagoons today, but since little is known about the early Earth, we can imagine
one there. To save space, I will call it Miller's lagoon.

But my reading led me to another hazard in the evaporative process.
Japanese chemists have reported that the simple amino acid, glycine, reacts
with urea more quickly than the chemical that forms cytosine. If glycine
were present in the lagoon, then, it would spoil the pudding. But would it
be there? The answer would definitely be yes, if we are stocking our ocean

by the product of lightning storms. Amino acids were the most prominent products of the Miller-Urey electric spark experiment. So interference by glycine appears a formidable roadblock, until prebiotic chemists devise some way to produce urea without glycine. We will note this as a problem to be solved by future work and return to Miller's lagoon.

We have concentrated our glycine-free urea solution in just a few years and are ready for the reaction. How much time will we need to carry it out? That will depend on the actual circumstances. If we assume a nice summer temperature of about 30°C (88°F), and fairly concentrated urea, it might take twenty-five years to get a good deal of reaction. This is much longer than the time permitted for rapid concentration. If our bathtub full of solution were spread over a large area, it would almost certainly dry up completely, killing the reaction. But a few modifications to the lagoon should alter that. We can give the lagoon a gentle slope draining toward the center, with a small crevasse or pocket to hold the final fluid. But now we need the other partner, to combine with the urea.

The partner's name, cyanoacetaldehyde, was furnished previously, but for the sake of the readers, I will call it CAT for short. The Miller and Robertson paper implied that CAT had been waiting patiently alongside the urea as the evaporation proceeded. In the *New York Times* article, CAT was called "another quite common component of sea water that also owes its formation partly to lightning bolts." But I knew of no reference documenting that fact.

My own search on CAT turned up some other alarming news about its behavior, however: It more resembles a prowling tiger than a sleeping kitten. It pounces avidly on almost every type of molecule that might inhabit a primordial soup: cyanide, amines, sulfur compounds, you name it. If deprived of any prey, it self-destructs. At the temperature we mentioned previously, half of it would be gone in thirty years (another reason why the evaporation has to be run quickly). For CAT to be common, some source must be pouring it into the oceans at a substantial rate. I turned to the *Nature* article for instruction.

This document only stated, however, that CAT came from the reaction of another substance, cyanoacetylene (I will call it CECIL), with water, and referred me to another paper. I learned that CAT is a tame pussy compared to CECIL. The latter can be produced in spark discharge experiments but reacts very rapidly with the other products. To demonstrate the production of CECIL at all, you need to use a special atmosphere unlike those proposed for the early Earth. We can flag this as another problem, but we will assume that CECIL got into the lagoon somehow, and cytosine was made. We can check off the successful preparation of another raw material needed for RNA world.

But we have barely begun the job of creating RNA. Our cytosine supply, which we have left sitting in a crevasse in a rock, must return to the sea so that it can meet the other components. Furthermore, this liaison must take place within a limited period of time. When it is left alone in water, cytosine self-destructs in a reaction that was discovered in my own laboratory some years ago. At room temperature, half of it will be gone in 300 years. To avoid this problem, we will specify that an earthquake ruptures the barrier and floods the lagoon basin soon after the cytosine has been prepared.

A batch of cytosine has been prepared and released into the sea in a warm temperate area. Other RNA components, according to the *New York Times* article, were prepared in polar regions. Each of them would diffuse out into the oceans that covered the planet, and there is a danger that the various pieces of RNA would simply get lost in that huge volume. We would need to produce much more cytosine if we need to stock the entire ocean with a sufficient concentration of it. But a calculation showed me that even if we lined all of the oceans of the Earth with Miller-type lagoons and had them churning out cytosine continually on a batchwise basis, we could not produce enough for that purpose.

Fortunately, other prebiotic chemists have tackled that problem. They have abolished the ocean model of the soup and replaced it with a plan for the early Earth in which the various tasks needed for RNA construction are placed in separate locations that are conveniently close to one another. One such illustration placed a glacier, a lava flow, hot springs, a pond formed by a soft comet impact, a lake named for Charles Darwin, and several other features in the same environment, as an aurora flickered overhead. Such a combination hardly seems common or likely, but we are compelled to adopt it if we maintain that pathways to RNA must have existed on the early Earth.

The Retreat from RNA

It is time for me to drop my role as an RNA believer, though other scientists continue to maintain the position. The intellectual appeal of the RNA world solution is so great that they are reluctant to abandon it. Jack Szostak and his colleagues acknowledge, for example, that "many prominent researchers in the field ridicule RNA as a plausible prebiotic molecule" but maintain that "our ignorance of prebiotic chemistry is severe, and it may be premature to dismiss prebiotic RNA entirely."

Szostak is certainly correct, in that we can only speak of pathways explored and not of those that may be yet unknown. But something definitive *can* be said about the entire field of prebiotic chemistry: It is much more an effort to dictate to nature, as we did with the medieval cosmos, than to learn from nature. And nature has been virtually waving its hands in the air and screaming at us. For an apt summary of the message, I shall turn to the British chemist, Graham Cairns-Smith, whom we shall hear from again shortly.

> There have been many interesting and detailed experiments in this area. But the importance of this work lies, to my mind, not in demonstrating how nucleotides could have formed on the primitive Earth, but in precisely the opposite: these experiments allow us to see, in much greater detail than would otherwise have been possible, just why prevital nucleic acids are highly implausible.

Nobel Laureate biologist Christian De Duve put the issue more succinctly when he asked, "Did God make RNA?" He proposed this in jest, but the remark brings us back to the division in philosophy that we came upon earlier. A spontaneous origin for RNA can be judged very implausible, but never impossible. If life started that way, we would expect life to be quite rare in the universe. This idea exactly fits the Sour Lemon outlook that we described earlier. Religious creation, as De Duve mentioned, is, of course, another choice. But we shall seek out others. It has been fashionable over the last decade, for example, for origin of life chemists to speak of "pre-RNA world."

I helped to launch this trend by pointing out the enormous difficulty in obtaining the sugar, ribose, in the prebiotic world. The case for ribose is even worse than the one for cytosine. Stanley Miller has contributed by measuring the stability of ribose. Even if it could be prepared successfully, half of it would decompose in 300 days at room temperature, as compared to 300 years for cytosine. For these reasons, many scientists have headed for the following escape hatch: Abandon ribose, but keep the idea of a replicator, so that natural selection can operate. Life then began with a simpler replicator, which functioned for a time in a pre-RNA world. At a certain point in evolution, it was replaced by RNA. The search for a suitable substitute replicator has become an exciting game, with several entrants in place.

Swiss chemist Albert Eschenmoser, who collaborated with my Ph.D. adviser on the total synthesis of vitamin B_{12}, found one surprising candidate. He had wanted to understand why ribose was the sugar of choice for the backbone of RNA, and also why ribose existed in a particular form, a five-membered ring, in RNA. Under other circumstances, ribose generally prefers to form an alternative six-membered ring. Using enormous energy and ingenuity, Eschenmoser and his co-workers constructed a number of RNA alternatives, including one that contained ribose in a six-membered ring. Surprisingly, the chemical properties of the six-ring version suggested that it would be a better replicator than the natural RNA. But it is hard to see why, once it got installed, it would ever yield to RNA. I would rather think that RNA assumed its present role accidentally, after functioning for some other purpose in an already existing cell. Professor Eschenmoser has been very reluctant to place any such strong interpretation on his results, though he has given his creation a name, p-RNA; I prefer the name "Swiss RNA" because of its greater efficiency.

Other alternatives exist. Stanley Miller prefers PNA, a combination of an amino acid backbone with the information units used by RNA. For this reason he has tried to sabotage ribose while rescuing cytosine. Others have suggested that even proteins themselves have an unrecognized ability to serve as the replicator. But I will argue that all replicators of this general type were very unlikely in the origin of life.

The Connection Problem, Revisited

Let's return to the monkey at the word processor and remember the calculation of the enormous odds that disfavored the production of the sentence, "To be, or not to be—that is the question." I used that example to question the likelihood of constructing a particular protein sequence at random. Despite the unfavorable outcome, the word processor contained a built-in bias that favored English sentences. More than half the keyboard was devoted to the characters used in the English language and relevant punctuation. Prebiotic chemists stack the deck even more strongly in their favor when they model the synthesis of a replicator. If they wish to connect

RNA building blocks, they include the appropriate components and exclude all others. The same selection process precedes the synthesis of any other proposed replicator. But nature will not be that kind.

Chemical mixtures made by natural processes will contain hordes of different components. We need only look at meteorite analyses (see Chapter 7, "Cosmic Sweepings") to get an idea of the complexity involved. When conditions force these molecules to combine with one another, they do not do so selectively. The nucleotides of RNA will not search each other out but may combine with nucleotides not present in RNA, fat components, amino acids, or any of the enormous number of the organic molecules not used in life. With one exception, these competitors have been rigorously excluded from the "prebiotic" syntheses carried out in the laboratory.

The exception involves a simple case. Amino acids, nucleotides, and many other molecules used in life have the property of "handedness." Like our right and left hands, they are not identical with their mirror images. If we want a typewriter analogy, then letters such as b, c, and d have handedness, while o and x do not. In experiments where mixtures containing both natural nucleotides and their unnatural mirror images were allowed to combine, both types were included in the product, spoiling its possible function.

For a more realistic illustration of the connection problem using a word processor, I would have to construct an enormous keyboard, with English characters intermixed with ones from Greek, Chinese, Arabic, and other languages, as well as nonsense symbols, numerals, and all manner of punctuation. A gorilla would be preferable to a monkey, as we would need a creature with enormous arms to work the keyboard. A typical gorilla product might be #*t5$\□⊃⇔φ⊕è∆✄📖6&Æq. I will not even try to calculate the odds of getting a string containing only English characters, let alone a meaningful sentence, in such a trial.

But even in this example, I am tilting the odds in our favor. Word processors will still align our typed characters neatly in a row. Nature is messier. For a better approximation, imagine the characters printed on square pieces, as in word games made with blocks of letters. Let us suppose that some pieces have a plug on one side and a socket on the other, so that they can be connected, just as a word processor connects letters. But now let us add an even larger number with only a plug or a socket. We can call them terminators, because they prevent the chain of characters from growing farther in one direction. Others with two plugs and no socket or two sockets and no plug will further wreak havoc in any effort to build regular sentences.

Most troublesome of all will be those with three connectors, which I will call branchers. If a gorilla were allowed to connect the squares, he would also start branches at random in the middle of existing lines. He would produce a two-dimensional tangle rather than tidy lines of characters.

Many building blocks of proteins and nucleic acids contain potential branchers, but branching is suppressed by the cell's machinery. Branching does occur when random mixtures of molecules combine and irregular netlike structures are produced. These structures could not function in any normal way as replicators.

Unless some guiding force were present (for example, a chemist), we could not expect anything useful to result from the connection of a mixture of components at random. Once again, there were no chemists on the early Earth. We cannot exclude the possibility that some mineral or other natural feature will selectively collect particular molecules from the mixture, but there is no reason to expect that the mineral would choose just the components that are needed for a replicator. We are asking the mineral to take over the purposeful role of the chemist. Of course, if we imagine that trials of this type are being conducted all over the universe, such an instance might turn up. Again, we find ourselves in the Sour Lemon position. But we do not as yet have to choose between the lemons and intelligent design. In the next chapter we will take up the Cosmic Evolution alternative, which requires no replicator. But first another possibility that is specialized but spectacular deserves to be mentioned. We will make use of it later.

The Way of Clay

The problems that surrounded us when we discussed prebiotic chemistry arose from the special properties of carbon. This atom is an essential construction unit for life as we know it on this planet. Carbon atoms happily connect to one another and to a variety of other atoms, and they can choose up to four partners at once. This versatility allows the construction of molecules with an extraordinary selection of properties, from the transparent plastics that we use to wrap foods to enzymes that can perform accurate molecular surgery on carefully specified clients. Our cells make broad use of the abilities of carbon but keep them under strict control to avoid chaos. But how does one avoid chaos in starting life?

Graham Cairns-Smith has made a startling suggestion, one that is heretical to orthodox biologists: Life began with a different chemical substance and avoided carbon compounds until the skills needed to handle them had been gained through evolution. Once carbon compounds were incorporated into the living process, their greater abilities soon asserted themselves, and the original materials were discarded. In his scheme, Cairns-Smith differs radically from the great majority of origin of life scientists, but he retains one feature that they cherish. He believes that life started with a replicator, but one that could arise without undue difficulty on the early Earth. Eventually, this primitive replicator was replaced by nucleic acids.

But what noncarbon substance could form a replicator? To introduce Cairns-Smith's choice, I shall steal the Creationists' thunder and cite the Bible: "But there went up a mist from the earth, and watered the whole face of the ground. And the Lord God formed man of the dust of the ground, and breathed into his nostrils the breath of life; and man became a living soul" (Genesis 2:6–7).

We encounter dust everywhere in many forms, but "the dust of the ground" probably signifies *clay*, a word often used to mean any finely divided mineral substance. Geologists more often use the word *clay* to signify a particular kind of mineral whose building block is silicate. The silicate group, like the carbon atom, can make four connections, but it is much more restrained in its choice of partners. Silicates are a major ingredient of the rocks of the Earth and often form regular structures in which a particular arrangement appears again and again. This monotony would appear to eliminate silicates as a choice for information storage, but Cairns-Smith feels that one subgroup furnishes an exception.

Cairns-Smith's selections grow in sheets, as do the shiny micas that we may find in nature. But the sheets that he nominates as the stuff of life are microscopic and accommodate imperfections in a way that stores information. They belong to a class called clay minerals, another use for the word *clay*. In the words of Cairns-Smith, "For a picture of first life do not think about cells, think instead about a kind of mud, an assemblage of clays actively crystallizing from solution."

A wonderful story can be created from Cairns-Smith's scientific speculations. We can picture the irregularities in the clay sheets leading to the formation of surfaces that have enzymelike properties. Such "clayzymes" might not have the exact molecular control of their modern protein counterparts, but the minerals could have other compensating advantages. They might be able to form their own microscopic chemical apparatus, with tubes, pores, pipes, and even pumps, all controlled by their sheets of mineral "genes." The sheets would grow by adding new layers at their ends that retain the information and properties. And when conditions are appropriate, they would reproduce, simply by splitting between the mineral sheets. Information storage in two dimensions and replication by splitting may not match the image that most origin of life chemists keep in mind, but the clay creatures would still qualify for the function they hold most dear: They could evolve through natural selection.

In our tale of clay, we can imagine three such creatures growing in cracks in porous rocks on the early Earth. One who we will call Tough grows in a way that blocks the flow of water. His attachment to the rocks is so firm that he enlarges but never splits. He may survive a long time but has no descendants and plays no further part in history. Another variant, Sloppy by name, allows water to flow by easily. He grows quickly. But when heavy rains fall and his crevasse is flooded, he disintegrates totally and ceases to exist. The third, and eventual winner, can be called Lumpy. He allows some water to flow through him and grows at a moderate rate. When the rains come, parts of Lumpy flake off, wash downstream, and start new growth. His descendants proliferate. In a very dry period, we can even imagine them forming dessicated spores and spreading by wind action.

Eventually, the clays will bring in carbon compounds to share the work, and modern evolution will begin. But the clays would be our earliest ancestors.

How much of this myth is based on demonstrated science, and how much on imagination? We can be secure that clay minerals were present on the early Earth; that is not true for any large carbon-based molecule with information-storing properties. Minerals possess irregularities; we do not know if the irregularities can control chemical reactions or be replicated. No evidence supports the idea that living beings can be made of clay, yet we cannot dismiss it out of hand. The media have picked it up from time to time, and the concept is respectable enough to be included as a debating position at origin of life conferences. It has not had enough clout, however, to merit funding on more than a token scale. Extensive studies would be needed to demonstrate the key properties in the laboratory; these have not taken place. Alternatively, we could search for such life, on Earth or elsewhere.

For now, clay life hangs in limbo. Unlike the case of the carbon-based replicators, we have not yet learned enough to decide whether they deserve Sour Lemon status. But we must move on, to ask whether we need replicators at all to get life started.

6

The Life Principle

A Dinner Out of Time

In *The Book of Lists*, the authors invite us to draw up a guest list for a dinner that we would like to attend, imagining that we may select any living or dead historical figure to share our table. Various celebrities submitted their lists, which included such notables as Gandhi, Benjamin Franklin, Jesus, Socrates, Pasteur, and so on. Using this model, I have imagined such a dinner on my own, but with one modification: Those invited share a particular interest in common, but they must discover what it is for themselves.

In this spirit, imagine that three guests arrive early at a table set for about a dozen. Two of them are dressed in the style of businessmen of the Victorian era, with suit and tie, vest, and a pocket watch on a chain. One of the two has a full, gray, bushy beard and speaks English with a German accent. He admits that he has played a role in his family business but his real interest is in revolution. He wants to see a just and progressive society established on Earth, and a strong controlling government will be needed to bring this about.

The second gentleman seems more nervous; he is bald but his face is surrounded by a fringe of gray whiskers. He is reluctant to speak but does so after some prodding. He has been trained as an engineer but has turned his attention to social philosophy. In contrast to the first, he supports extreme economic and social freedom for the individual, and limited government. Before the two can quarrel, a third guest arrives, in clerical attire. He

123

is clean shaven, with gray hair and a lean face. He speaks English with a French accent. He discloses that he is a practicing Jesuit but also has had a lifelong interest in paleontology. His dinner companions, who have abandoned the Christian faith, show no empathy for this calling.

The beginning has been unpromising, but after further discussion, the three find that they hold close views on how the universe, life, and human society had come to its present situation. All believed in progress and evolution on the cosmic scale.

The revolutionary, Friedrich Engels (1820–1895), was the closest collaborator of Karl Marx and coauthor of *The Communist Manifesto*. In his later book, *The Dialectics of Nature*, Engels extended their principle of dialectical materialism to the natural world. He argued that nature exists in a series of levels, in which new principles of complexity or order emerge through time. The principles at one level cannot be understood in terms of laws operating at lower levels.

The philosopher, Herbert Spencer (1820–1903), vigorously endorsed the idea of evolution, in biology and in the cosmos. In fact, he had first coined the term *survival of the fittest*. He believed that all aspects of reality involved a continuing development from lower stages to higher ones. An unknown and unknowable force triggered this development, producing order from disorder.

The cleric, Pierre Teilhard de Chardin (1881–1955), had attempted to synthesize a philosophy that combined Christian and evolutionary concepts. As the Church did not approve of his ideas, his book, *The Phenomenon of Man*, could not be published until after his death. In his view, cosmic evolution unfolded as a slow, gradual process until it reached a sudden discontinuity, or critical point, as water does when it reaches the boiling point. The process was driven by an all-pervasive ascending force, sometimes called "complexification."

Although their philosophies of the past were similar, the three disagreed about the course of the future. Engels felt that life and humanity resulted from a continuous evolution of matter. At a much higher level, these same forces directed societies to socialism. To Teilhard, evolution of humans would now continue in the realm of the mind, called the "noosphere." Humans would converge, not diversify, until they reached a final, almost divine condition called the "Omega point." Spencer did not espouse a detailed plan for further evolution.

A second table has been set for dinner in the hall, within earshot of the first. Again, the table holds individuals from different backgrounds and eras. As the conversation drifts over from the first table, however, the guests at Table 2 find their own unifying theme: They dislike the ideas of the other group. The French Nobel Prize-winning biologist, Jacques Monod, sits at the head of the table. We have met him before, as a key member of our Sour Lemon School. He feels that Engels's ideas conflict with modern science and voices these opinions about those of the cleric:

> Although Teilhard's logic is hazy and his style laborious, some of those who do not accept his ideology yet allow it a certain poetic grandeur. For my part I am most of all struck by the intellectual spinelessness of this philosophy. In it I see more than anything else a systematic trucking, a willingness to conciliate at any price, to come to any compromise.

Another arrival at the second table is the contemporary Harvard biologist, Stephen Jay Gould. He is an expert on matters of evolution and the author of many widely appreciated books on the subject, but his studies have led him to very different conclusions: "Progress is a noxious, culturally embedded, untestable, nonoperational, intractable idea that must be replaced if we wish to understand the patterns of history."

But his following comments reveal him to be partly in the other camp. He dislikes the concept of an organizing force in evolution but accepts it in the origin of life:

> The discovery of evidence that life may exist elsewhere in the universe raises the most profound of all human questions: why does life exist at all? Is it simply that if enough cosmic elements slop together for enough eons, eventually a molecule will form somewhere, or many somewheres, that will replicate itself over and over until it evolves into a creature that can scratch its head? Or did an all-powerful god set into motion an unfathomable process in order to give warmth and meaning to a universe that would be otherwise cold and meaningless?

Gould suggests a third alternative:

> Suppose that the simplest kind of cellular life arises as a predictable result of organic chemistry and the physics of self-organizing systems whenever planets exist with the right constituents and conditions—

undoubtedly a common occurrence in our vast universe. But suppose, in addition, that no predictable directions exist for life's later development from these basic beginnings. . . life of bacterial grade may arise almost everywhere and then proceed nowhere in particular, if anywhere at all—a perfectly splendid outcome.

The next guest to join the second table is Sir Fred Hoyle, the much honored British astronomer. He sets out immediately to contradict Gould on the origin of life:

If there were some deep principle that drove organic (carbon-rich) systems toward living systems, the operation of the principle should be demonstrable in a test tube in half a morning. Needless to say, no such demonstration has ever been given. Nothing happens when organic materials are subjected to the usual prescription of showers of electrical sparks or drenched in ultraviolet light, except the eventual production of a tarry sludge.

The next guests who arrive at the second table also dislike the philosophy at the first one but operate from different presumptions. A balding, clean-shaven, stocky individual joins the table and announces with a rhetorical flourish, "Chemistry refutes all the claims of the evolutionists, and proves that there is no pushing power to be found anywhere in nature—no progressive force at work in the earth—no eternal urge lifting matter or life from any plane to a higher one." This gentleman is William Jennings Bryan, three-time Democratic candidate for President near the turn of the twentieth century (he lost all three elections). He is joined by Henry Morris, organizer of the Institute for Creation Research, who adds his voice to the consensus that denies Cosmic Evolution:

Where in all the universe does one find a plan which sets forth how to organize random particles into particular people? And where does one see a marvelous motor which converts the continual flow of solar radiant energy bathing the earth into the work of building chemical elements into replicating cellular systems, or of organizing populations of worms into populations of men, over vast spans of geological time? . . . The answer is that no such code and mechanism have ever been identified.

The Reinforcements Arrive

As the dispute heats up, the group at the first table is increasingly outnumbered and outshouted. They learn from their opponents that Spencer's philosophy is quite out of fashion in the twentieth century, that Communism has been tried in many countries and lies in ruin everywhere, and that Teilhard's ideas have acquired a certain cult following but never wide acceptance. At this discouraging point, however, they are reinforced by the arrival of a host of newcomers from the present time, most of them winners of the Nobel Prize. These Nobel Laureates include physicists Philip Anderson, Ilya Prigogine, Manfred Eigen, and Murray Gell-Mann; biologist Christian De Duve; and economist Kenneth Arrow.

The Nobel Laureates have all endorsed at least a part of the ideas of the gentlemen at the first table and given them current life in a new area called "complexity theory." To explore the applications of complexity, the Santa Fe Institute was founded in 1984 and now has a budget of about $5 million per year. Ironically, many of the Laureates had been inspired by the work of Monod and François Jacob on genetic regulation, which they saw as one of the applications of this theory.

This new area has not embedded its ideas in a socialist, libertarian, or theological setting, yet many of their concepts remain the same as those voiced by earlier philosophers. Their treatment is mathematical and lengthy but contains the following message: They consider situations in which a group of players or "agents" have many interlocking interactions with one another. The agents may be buyers and sellers in the stock market, interacting genes, reactive chemicals in a mixture, neurons in a brain cell, or winds in the atmosphere. The type of system that results is classified into one of three areas:

1. Simple systems that consist of a few agents. The interactions are limited and governed by well-understood laws. They can be described in equations and their behavior predicted. One example involves the motions of the Earth, Moon, and Sun about one another. Some of my friends have planned vacations in areas where solar eclipses are expected, and they have never been disappointed. Eclipses are predicted centuries in advance, and they arrive in timely fashion, to the minute.

2. Chaotic behavior. You can place yourself under a solar eclipse, but you cannot predict that the cloud pattern will allow you to see it. Years ago, a group of friends and I traveled to Maine to watch a total eclipse that

would cut a path across the state. We saw amateur astronomers setting up telescopes in open fields at several locations. But we had a different strategy. The sky displayed a mixture of mottled clouds with patches of sun. The forecast said nothing about the locations where the sun might be visible during the period of full totality but only promised "mixed clouds and sun." So the group of us piled into our rented car. One of us drove, another scanned the sky for open patches, and a third peered at a roadmap to locate a suitable road pointed in the right direction. The strategy worked—we broke out into sunlight ten minutes before totality and watched the glorious sight. Most of the fixed-position astronomers were not so lucky. Some observers had small planes prepared at the airport but could not get them off the ground in time when clouds suddenly covered the sky just before the crucial moments.

Some aspects of weather forecasting are inherently unpredictable in detail. They follow behavior that is described as chaotic. When scientists attempted to model such systems in detail, they found that, in the words of science writer James Gleick, "Tiny differences in input could quickly become overwhelming differences in output—a phenomenon given the name 'sensitive dependence on initial conditions'. In weather, for example, this translates into what is only half-jokingly known as the Butterfly Effect—the notion that a butterfly stirring the air today in Peking can transform storm systems next month in New York."

3. Complexity. Between the realms of simple order and chaos lies another region, that of complexity. It is a concept that can be grasped through examples more readily than it can be defined precisely. The repetitive sound of a beeper illustrates simple order, and the noise of traffic at a busy street intersection is chaotic. A symphony, however, is complex. Another description can be found in the words of the late physicist Heinz Pagels: "A rose, which has both randomness and order in the arrangement of its parts, is 'complex'; the movement of gas molecules is truly 'chaotic'. Complexity thus covers a vast territory that lies between order and chaos."

We will adopt Pagels's term, though others, such as "self-organizing criticality," have also been used. To enter the land of complexity, you must have the right number of players, with a suitable amount of interactions connecting them. If you reduce or enlarge these variables beyond certain crucial limits, then you settle into order or collapse into chaos. The development of modern high-speed computers was essential for this field, to permit researchers to keep track of the behavior of the multiple players, and to follow the fate of the system. One of the best-known examples that has been studied in this way is the sand pile model of physicist Per Bak.

Dreams Built of Sand

Imagine a level round table with a flat surface, onto which we spill a steady flow of some material from a higher point. We know the rate of flow from above, and we can measure the rate at which material hits the floor, after spilling off the table. If the material were water and the table made of glass, our system would be simple. The water would pour off the table at the same rate that we added it from above. Everyone but the person who has to mop up the floor would find the system boring. When we substitute sand for water, however, things become more interesting.

For a time, the sand will pile up on the table, as resistance to flow holds the grains together. Finally, the mound reaches a critical height, and sand starts to spill off the table onto the floor. But even though our addition rate is steady, sand does not reach the floor in a constant flow. Instead, a series of small and large avalanches occurs, with periods in between where no sand falls off. A single grain added at a critical moment can trigger a large event of this type.

So far the situation seems chaotic. We cannot predict when the next avalanche will take place and how large it will be. But if we studied the pile of sand for a long enough time, we would see that the distribution of the size of the slides follows a regular pattern that is known as a power law.

To do this, we would measure the size of all of the avalanches and divide them into groups by size. For example, the smallest group might contain 0 to 100 grains, the next 100 to 200, and so on. We would count how many avalanches belonged to each group. We would then take the logarithm of the size of the avalanches in each group and the logarithm of the number of members of that group. These numbers would be placed as a point on a graph. When we looked at the overall collection of points, we would see that they formed a straight line. If you plot the logarithm of the size of an avalanche against the logarithm of the frequency at which it takes place, you get a straight line. This relation means that for every million avalanches that you got of magnitude 1, you would get 100,000 of magnitude 10, but only 1 of magnitude 1 million. Obviously, if the larger, most catastrophic ones were most interesting to us, we would find computer modeling to be the most effective way to study the system. Otherwise, we might have to wait quite awhile for a large one.

The occurrence of catastrophic avalanches is called an "emergent" property in complexity theory. We could not learn that such events will happen by studying the properties of individual sand grains. The formation of a whirlpool, as water flows steadily down a drain, has been cited as another emergent property. The flow of energy through the system, in these cases the gravitational energy of the falling sand or the flowing water, leads to the appearance of such new properties.

Although the sand pile experiment can be carried out most readily on a computer, some effort has been made by a Norwegian group and others to study the system experimentally. They found that spherical sand particles did not follow the computer model well, but that they could approximate the model by switching to piles of elongated grains of rice. The longer grains increased the friction, and this change was needed for the system to follow the rules of complexity. When the shape of the stationary rice pile was examined, it showed a fractal pattern—one in which the surface pattern is repeated at ever smaller scales. Individual colored grains were followed, and their progress was quite complicated. Some grains remained in the pile for a very long time before falling off the table.

Other laboratory studies have found that self-organization can appear in surprising circumstances, for example, in trays of tiny brass balls, when the balls are jiggled up and down. At certain vibration levels, ripples and other features form. Square and stripe patterns may pulse up and down. At lower frequencies, isolated peaks and valleys form. If the brass "sand" is touched with a pencil, a depression is formed that drifts around. As it does so, it oscillates in shape between a crater and a peak. Some chemical reactions also set up oscillations, sometimes accompanied by brilliant color phenomena. A reaction of this type was described in *Scientific American:* "The solution is initially a uniform purple. . . . As the reaction proceeds white dots appear and grow into rings and sets of concentric rings which annihilate each other when they collide." One observer compared the emergence of white dots against the purple background to seeing "stars come out." In other reactions, beautiful patterns of alternating red and blue rings appear.

Does this behavior have any importance in the larger world? According to Per Bak, "Earthquakes may be the cleanest and most direct example of a self-organized critical phenomenon in nature." The Earth's crust remains at rest most of the time, but there are intermittent periods of violent activity. The frequency and size of earthquakes follow a power law, with many smaller quakes and a few large ones. In addition, the faults form fractal patterns.

Others have attempted to find an analogy between the sand pile and the evolution of life, which has also been punctuated by massive extinctions and spurts in the appearance of new species. In his own book, Bak quotes Vice President Al Gore: "The sand pile theory—self organized criticality—is irresistible as a metaphor; one can begin by applying it to the developmental stages of human life. . . . One reason that I am drawn to this theory is that it has helped me understand change in my own life."

Complexity scientists consider their emerging discipline not just as a metaphor, but one with predictive power in a vast array of natural and human situations, from commodities futures and traffic jams to hurricanes. The basic concepts dovetail well with those of Cosmic Evolution.

But of course, the collapse of a sand pile is not enough to suggest the massive increase in organization that characterizes the development of living systems. For a proper analogy, we would have to see the sand pile gradually develop into a sand castle, complete with bridges, towers, and windows. I suspect this will not occur, no matter how long we run the sand. But other aspects of complexity theory predict that this is just what can take place when a more organized system operates long enough to reach a critical state.

At that point, a dramatic change can take place and the system moves up to a higher level of organization. This has been called a "phase change," analogous to the abrupt change when water freezes. The agents at one level act as building blocks at the next level, which differs in ways that could not be predicted from study of the previous one. As Nobel Laureate Philip Anderson wrote, "at each level of complexity entirely new properties appear. . . . At each stage entirely new laws, concepts and generalizations are necessary. . . . In fact, the more the elementary particle physicists tell us about the nature of the fundamental laws, the less relevance they seem to have to the very real problems of the rest of science, much less society."

In this way, a complex adapting system gradually develops multiple layers. In the case of living systems, simpler molecules give rise to proteins and other large molecules, which in turn assemble into organelles. At higher levels we find tissues, organs, living creatures, and entire ecologies. Many components of our economy and government are multilayered and can be fit to the same pattern. The layered structure that exists today mirrors the historical process of self-organization that gave rise to it.

The Hecklers Persist

In spite of the endorsement of its ideas by many distinguished scientists, complexity theory has not obtained universal recognition, or even respect. Let us return to our Dinner Out of Time and imagine that additional guests now arrive to fill in the vacant seats at the second table. Science writer John Horgan enters the fray, pointing out the "gap between rhetoric and reality" in this field. In discussing Per Bak's contribution, he adds,

> Self-organized criticality is not really a theory at all. . . . Self-organized criticality is merely a description, one of the many, of the random fluctuations, the noise, permeating nature. By Bak's own admission, his model can generate neither specific predictions about nature nor meaningful insights. What good is it, then?

British geneticist Gabriel A. Dover also joins in, with particular reference to the ideas of Stuart Kauffman (a biologist at the Santa Fe Institute, and a guest at the first table):

> There are times when the bracing walk through hyperspace seems unfazed by the nagging demands of reality. Is there truly a "beautiful order which graces the living world" requiring recourse to self-organized properties of large numbers of interacting units? . . . In defence of the concept of life as the cookie crumbled, I would answer no.

Mathematician David Berlinski then sums it up: "I find nothing of value in various theories of self-organization; the very idea is to my mind incoherent."

The Origin of Life Super Bowl

I could extend my imaginary dinner debate late into the night without appreciable progress. In real life, the debate has continued in this way for well over a century, with or without computer models. In science, as in the

Super Bowl, championships are not decided by invective. Instead we proceed to the critical test in a selected arena. As I have suggested in earlier chapters, the ideal playing field for the complexity and Cosmic Evolution questions is the origin of life. The Creationists and the advocates of intelligent design feel that this problem can never be solved by science, as supernatural means were used for the purpose.

To members of the group that I have called the Sour Lemon School, the origin of life was an extremely improbable and unique event. By studying present life, we can infer something about the circumstances, and perhaps suggest a series of unlikely happenings, but we can never be certain about the details. They have vanished into history. Experiments that attempt to model them usually require massive intervention by the scientist and thus demonstrate a version of intelligent design.

For complexity theorists, however, a demonstration of the start of life should be possible. Biologist Stuart Kauffman has been the leading advocate of this idea.

In his computer models, Kauffman uses a group of interacting catalysts as the agents in a complex system. Some of them react directly with one another. Others act indirectly, by promoting reactions in which they do not participate. They act as protein enzymes do, but more weakly. When the number of chemicals and interactions has attained a critical value, an "autocatalytic set" is formed, which grows at the expense of the unorganized chemicals around it. After the set has reached an appropriate size, it divides, giving rise to smaller "daughters." Reproduction has taken place without the special mechanism of replication (exact copying of a large information-bearing molecule). I would like to flesh out this scenario further, by using Kauffman's own words, culled from his book *At Home in the Universe*:

> Order is not at all accidental. . . . Vast veins of spontaneous order lie at hand. Laws of complexity spontaneously generate much of the order of the natural world.

> Life is a natural property of complex chemical systems. . . . When the number of different molecules in a chemical soup passes a certain threshold, a self-sustaining network of reactions—an autocatalytic metabolism—will suddenly appear.

> Autocatalytic metabolisms arose in the primal waters spontaneously, built from a random conglomeration of whatever happened to be

around. One would think that such a haphazard collection of thousands of molecular species would most likely behave in a manner that was disorderly and unstable. In fact, the opposite is true: order arises spontaneously, order for free.

The collective system does possess a stunning property not possessed by any of its parts. It is able to reproduce and to evolve. The collective system is alive. Its parts are just chemicals.

Once an autocatalytic set is enclosed in a spatial compartment of some sort . . . the self-sustaining metabolic processes can actually increase the number of copies of each type of molecule in the system. In principle, when the total has doubled, the compartmentalized system can "divide" into two daughters. Self-reproduction can occur.

If all this is true, life is vastly more probable than we have supposed. Not only are we at home in the universe, but we are far more likely to share it with unknown companions.

As written, the preceding statements are philosophy. They need to be demonstrated to enter the realm of science. Science writer M. Mitchell Waldrop has described the evolution of Kauffman's ideas concerning such a demonstration:

No matter how many calculations and computer simulations he carried out on the origin of life, they were still just calculations and computer simulations. To make a really compelling case, he would have had to take the experiments of Miller and Urey one step farther, by demonstrating that their primordial soup could actually give rise to an autocatalytic set in the laboratory. But Kauffman had no idea how to do that. Even if he had had the patience and knowledge to do laboratory chemistry, he would have had to look at millions of possible compounds in all conceivable combinations under a wide range of temperatures and pressures. He could have spent a lifetime on the problem and gotten nowhere. No one else seemed to have any good ideas either.

Years passed, and the experimental situation did not improve, but computers got much more powerful and versatile. As reported by Waldrop,

Kauffman was thinking about his data one day while walking in the Sierras, "and suddenly, I knew that God had revealed to me a part of how his universe works." This comment was not a reference to a personal God, but rather a reference to the feeling that came over him as knowledge was unfolding: "It was a lovely moment, the closest I've had to a religious experience."

According to Kauffman, he had modeled his reaction network by using on and off "light bulbs" on the computer. The condition of each bulb was determined by those of other bulbs, according to a set of rules, which included some probability factors. The fate of the system depended on the number of bulbs and interactions of various types. The system could lock into a simple state of order, particularly if there were few agents. If a single bulb had too large a number of interactions (which I suspect is the fate of most real mixtures of organic chemicals), the system wandered into chaotic nonrepetitive behavior. If the interactions were set at an intermediate value (2 seemed optimal), an autocatalytic set could form.

When I first read this account, it reminded me of a game that I had acquired for my own computer. It had arrived as one component of an entertainment package sold by Microsoft. The documentation informed me that this entertainment was a descendant of a Game of Life invented by John Conway and first described in the October 1970 issue of *Scientific American*. The 1970 game was followed by pencil and paper, but my computer version runs with much less personal labor. At the start I see a grid of squares before me. Squares can be either empty (dead) or else colored red or blue (alive). The colors represent two competing species. The board then is recalculated to represent the passing of a generation. Whether a particular "living cell" or square survives or not depends on the number of neighbors that it has. If there are too few or too many, the cell dies of loneliness or suffocation. Otherwise it survives. Another rule decides when a dead square springs back to life again.

I set the game to run automatically and watched as shifting patterns of colored cells flashed to mark the passing of generations. The pattern altered constantly, but after many generations had come and gone, it settled down to an alternation of two simple patterns. No further change took place; evolution in this case had reached a dead end. The documentation told me that in some cases more than a thousand generations could pass before the pattern settled down into a repetitive mode.

Kauffman's simulation, like the Game of Life, functions entirely in the realm of the computer. To relate it to the actual origin of life, we would need to find a real system of chemicals that behaved in this way and continued to evolve. The initial assessment described by Waldrop still held.

In his writings, Kauffman has suggested that a mixture of small protein enzymes, with a moderate number of linked subunits, might behave suitably, but no experiments have been carried out. For the reasons that I put forward in the last chapter, I think it very unlikely that enzymes would form spontaneously in a chaotic mixture.

Other groups have suggested different chemical combinations. For example, Doron Lancet and his collaborators at the Weizmann Institute of Science in Rehovot, Israel, have drawn their inspiration from the membrane that surrounds cells, rather than the enzymes within it. In the same way that droplets of oil seek each other out in the presence of water, certain lipid molecules will assemble to form a variety of very different microscopic compartments. Some of these compartments may have the ability to recruit or make additional molecules of the same types. They will grow larger and eventually split into two daughters. These offspring will, in a primitive way, have inherited information from the parent compartment. This scenario seems chemically more reasonable for the early Earth than does the formation of small enzymes made only of amino acids, but again, it remains a concept that only inhabits the computer. What has been lacking is a convincing experimental demonstration.

In fact, one promising scheme has emerged in the last decade that appears to embrace some of the key concepts of complexity theory. It requires no replicator—a set of simple small molecules carries out cycles of reactions that gradually evolve to greater complexity. The specific processes involved have been spelled out in considerable detail, and several of the novel ones have been demonstrated to work in the laboratory. When one particularly striking reaction from this group was reported in the *New York Times* one day in the spring of 1997, I found myself drafted for an appearance on a national cable television newscast that same afternoon. I had made no contribution to the work, but there is a shortage of origin of life scholars in the New York City area. Leslie Orgel of the Salk Institute, one of the most profound thinkers in the field, later told me that he considered the work to be the most important finding in the origin of life in the last half century.

The originator of this novel concept is a German patent attorney named Günter Wächtershäuser. He chose his profession after obtaining a doctoral degree in chemistry. On those occasions when I heard him speak at origin of life conferences, I felt that he was combining these skills, as he advanced his ideas with the precise logic and manner of a prosecuting attorney.

He came to his theories by long nighttime study of the materials that he felt to be the most relevant: the biochemical cycles that work in life today. He deduced that certain central processes were likely to have been around at the start, and then he asked *what else* would have been needed to make them function. Perhaps some key supporting player was present that had since been discarded, in the way that Cairns-Smith's clay genes were tossed aside when organic replicators appeared.

Wächtershäuser's selection, like that of Cairns-Smith, was a mineral. But he did not propose it as the sole stuff of life, or that it acted as a genetic material. In his proposal, the mineral served as a source of the necessary energy and also as a staging arena. Many of the key players in the central cycles of life today bear negative electrical charges. Wächtershäuser proposed that his mineral had a positive surface. It could then select certain chemicals from the chaotic mix that prevailed in the surrounding waters and offer its surface as a stage on which the chemicals could encounter each other and do their business. Of course, he had to identify a mineral that had the necessary properties for the part he assigned to it. His choice was an iron sulfide called pyrite, whose shiny chemical luster had led others to call it "fool's gold." The unsavory background of his chosen substance has not fazed Wächtershäuser: "Pyrite is frequently encountered by gold diggers. It glitters like gold. This is the reason why it is called fool's gold. But I would like to predict that once the secrets of pyrite are known, only a fool will mistake this mineral with something as base as gold."

It was not easy, of course, for an outsider, a scientist with another profession and no laboratory, to get a hearing for his theories. Fortunately, Wächtershäuser had gained the early support of evolutionary biologist Carl Woese and philosopher Karl Popper. With their backing, his amazingly detailed proposed schemes of reactions could find their way into print. When I examined his longest paper, I felt as if someone had handed me a biochemistry text from the late twenty-first century. Most of the chemicals were familiar, but they were organized into an evolutionary pattern that I had never seen before.

Wächtershäuser's exposed position leaves him vulnerable, in that his proposed reactions can be tested and shown to fail. He has, however, attracted some experimental collaborators, and they have made a few of the steps work. He has not made a dent, however, on many key supporters of the replicator idea, whose very definition of life excludes reaction cycles of this type from consideration as a living system. They would rather search for improved ribozymes.

Wächtershäuser, on the other hand, has pressed the exploration of his own specific system but has not attempted to link it with complexity theory, though that would seem to me to be a likely connection. He continues to demonstrate one reaction of his scheme at a time, while those who believe in RNA world belittle the significance of his demonstrations. So how can a solution emerge from this impasse?

In an earlier book I proposed that workers of all persuasions run prebiotic simulations rather than "prebiotic" syntheses. We have already seen how the scientists following the latter approach have kept quite busy in attempting to demonstrate one or another chemical process that they felt took place on the early Earth. On reading their papers, I often remembered a skilled lion tamer whom I had once seen in the circus. This individual got the beasts to sit on stools, leap through rings, and even refrain from biting his head off when he put it between their jaws. It was a stunning demonstration, but I did not imagine that this was what the animals chose to do on their own, when left alone in the jungle. To learn that, we would do better by setting up unseen cameras in the wilderness.

By analogy, we might learn something about how life started by just observing the beasts instead of putting our whip to them. In a prebiotic simulation, we would select a likely group of chemicals that we felt might have come together at the start of life and provide them with a plausible energy supply. For example, we could choose the Miller-Urey product and expose it to simulated or real sunlight, or select the Wächtershäuser mix with an automatically replenished supply of pyrite. We would construct an appropriate setting, which simulated a pond, a hot spring, or any other favored environment. The key point is that once we had started a run, we would not interfere further, except to withdraw samples to be analyzed. We would simply watch what happened.

In some cases, a simple kind of order might emerge, the unchanging condition that chemists call equilibrium. In other cases we might veer into chaos: the random formation of chemical bonds leading to a messy (and hard to clean out!) tar. But just possibly, we might hit a system that started to evolve and did not stop. If so, we would then be able to watch self-organization at work in the real world rather than in a computer. We might be exploring our own roots, repeating steps that took place on this planet billions of years ago. Alternatively, we may have hit a set of conditions that never applied here but worked to start life on some different world. In either case, we would be watching a general principle at work and studying its boundaries. This process deserves its own name, even though it falls within the larger area of complexity theory. The one that I suggested earlier was the Life Principle.

There has been no rush of chemists attempting to try a prebiotic simulation. The chance of failure is great, particularly in the first efforts. All of the difficulties that were set out previously, in the paraphrase of Kauffman's thinking provided by Waldrop, still apply. We have no guarantee that the Life Principle exists. If it does, it might not be possible to demonstrate it within the space and time limitations that a laboratory imposes. It is hoped that some ambitious, energetic, and young scientists will take on the effort. But I have written this book to propose an additional approach to the problem. Nature, anticipating our curiosity, has been running experiments of this type on the grandest scale for billions of years. We and the other life forms on this planet represent one result. We now need to take a careful look and see what else has emerged.

Decision in the Solar System

Through most of human history, we stared at the planets with our naked eyes and used the planets as the building material for our dreams. At first these dreams involved Earth-like landscapes and creatures much like those we knew on this planet. Improved observations, first by telescope and more recently by passing spacecraft, have dissipated these dreams. In the case of the Moon, direct human presence confirmed what we had recognized earlier: that it is an alien inhospitable world bearing little resemblance to our own. No Earth-like worlds will be found in the remainder of the solar system either, but we may find out there the answer to one great question that weighs heavily in our cosmology: Is life a unique phenomenon, limited perhaps only to the Earth, or is it "written into" the laws of the universe?

Two philosophies, both rooted in science, differ profoundly in their prediction as to what we will encounter. The one that I have termed the Sour Lemon School predicts that the remainder of the solar system should be as barren as the surface of the Moon. We could expect little better in the remainder of the universe. As philosopher Paul Davies put it, such pessimists "believe that life is a freak—the result of a zillion-to-one accidental concatenation of molecules. It follows that the likelihood of it happening again elsewhere in the cosmos is accidental. . . . Life bucks this trend only because it is a statistical fluke."

This view represents a reaction against traditional religion and against the fanciful tales of creatures on other worlds that were accepted by many intellectuals for centuries. It goes beyond our present knowledge of other worlds to paint an utterly depressing picture of a barren and hostile universe: one that is unlikely to inspire us to explore it. Its weakest link is a resort to a random improbability to explain the most important feature in the universe: our own existence.

Complexity theory and its component, which I have called the Life Principle, offer a very different view. The universe is pictured as a vast incubator with a built-in tendency to create and nurture life whenever certain requirements are satisfied. Biologists have been much more conservative, assuming that life elsewhere must be like our own, and that only Earth-like planets can support it. As a result, astronomers have written about "habitability zones," and scientists have designed tests for earthly bacteria and assumed that they were also appropriate for extraterrestrials. In fact, our ignorance in this area is total. We have nothing in our experience to guide us concerning what other systems might serve as a basis for life, or how the creatures would function.

Some years ago my friend and coauthor, the late physicist Gerald Feinberg, tried to list a *minimal* set of requirements for chemically based life forms (he was also open to the possibility that exotic creatures might operate on a very different physical basis). In his theory, self-organization was possible whenever an appropriate set of simple chemicals, an energy supply that interacted with them, and a liquid or dense gas medium were put together. If the process were allowed to continue long enough, then enough organization could build up for us to call it life.

We know the recipe that worked on Earth. The chemicals probably contained carbon, water was the medium, and the ultimate energy source was the Sun. Yet if a robot intelligence that was ignorant of our type of life was shown this recipe, it would probably conclude that it would not work. It might feel that the tendency toward chaos would be overwhelming because of the excessive number of compounds that can be produced by carbon chemistry. If the robot was then presented with the results of analyses of meteorites, it might feel that its conclusions had been validated (we will get more deeply into this topic in the next chapter). A bewildering array of small carbon compounds can be found there, together with the amorphous, tarry materials that are formed when the small compounds combine at random. The robot might not understand how these chaotic tendencies could be restrained. Given our ignorance of the origin of life, we do not understand it as yet either, but obviously the problem has been solved in nature. When we seek out alien life, then, we must not let our natural carbon chauvinism blind us and cause us to miss what may be out there.

We can use these requirements to narrow our choice of sites. Chemical reactions run well in liquids and dense gases, but poorly in solids where the players are immobilized. The same considerations can apply in human affairs; if you are looking for a congenial companion, your chances are much better at a party than at a dinner table where you can only speak to the people seated close to you. We know that water works in forming life, so places with liquid water are a prime target. But some worlds have other liquids, and we will want to look at them, too. On Saturn's moon Titan, for example, we may discover lakes or seas full of liquified natural gas. A horrible cold envelops them, one that would bring our familiar biochemical reactions to a grinding halt, but other sets of reactions might find the temperature and the setting blissful.

On the other hand, the solar system contains a number of solid, small, airless worlds that fail to pass muster. We would not want to look at them first if we were trying to catch the Life Principle in action. But as we shall see in the next chapter, many other places still qualify. They have been around for over 4 billion years, as they and the Earth were all formed at about the same time. They may or may not have met the other requirements needed for self-organization. The simplest way to find out what has happened will be to go there and look around.

Before we start a tour in our imaginations, to flag the most promising worlds for our real-life quest, we will want to consider what we will gain if we succeed.

A Universe Full of Life

If our search turns out well, then we locate life forms that are very different from our own somewhere else in the solar system, as well as other evidence that the Life Principle is busily at work. From these examples, we will also gather some clues about the way that our own life arose, even if all the details are not resolved. Our origin will no longer be a profound mystery. Biology would then become a many-dimensioned discipline—we would have a symphony rather than a single note. As Carl Sagan has written, "The discovery of a single extraterrestrial organism—even something as humble as a bacterium—would revolutionize our understanding of living things."

The search should not end with that first discovery. We would be encouraged to explore farther—to see what other wonders have been produced over billions of years by the Life Principle. If we wandered farther, we might encounter other intelligence, or else learn that we are precocious and among the first class to "graduate" to that level. Or we might learn that evolution on other worlds has produced results that are as surprising and complex as intelligence, but very different in nature. A universe full of wonders may await us.

I find these possibilities far more satisfying than one in which the cosmos represents an indifferent and sterile pile of debris. A new story for our times will be shaped in which we are an important product of Cosmic Evolution, though not the only one. But what would happen to the older ones?

The discovery of a single instance of life in our system with a separate origin would shatter the Sour Lemon position. Some diehards could argue that we were even luckier and that two grand prize lottery winners, by chance, happened to dwell on the same block, but most of us would be happy to wager that something more than chance was at work.

The position of religion would be much more interesting, and we could expect many different reactions. I have heard one Fundamentalist response already, on my visit to the Institute for Creation Research. According to one guide, if life were discovered on Mars or Europa, the Fundamentalists would assume that it was created as well. For the believers in a biblical cosmos, business can go on as usual.

But options for more creative responses do exist. I am an agnostic, but if I were inclined toward religion, I know which of the two alternatives I would prefer. A fertile universe attracts me much more than a barren cosmos that requires divine intervention if it is to bear any fruit at all. In the former case, the Creator could set the general pattern (I have borrowed this phrasing from William Whewell) and then watch His handiwork evolve on a vast scale, turning His attention to the most successful locales when they matured.

Some philosophers have moved toward this position ahead of the needed evidence. They have drawn encouragement from the Anthropic Principle, a viewpoint that claims that our existence depends on an extraordinary series of coincidences. If the values of the basic physical constants that govern our universe were to change even slightly, then life could not have arisen here, according to these philosophers.

Such arguments are hardly secure, as our ability to calculate the properties of entire universes from their fundamental constants is not well developed. Yet the supposed coincidences have been used to infer the existence of a Creator. As Patrick Glynn pointed out in the *National Review,* "The Anthropic Principle does not settle the question; it is not a proof of God. But it alters the presumption; it shifts the *burden* of proof." But he is also quick to point out, "The God of the Anthropic Principle is not yet the God of Abraham, Isaac and Jacob."

The existence of a Life Principle would provide a more secure platform for those who wish to make such an inference than the Anthropic Principle. But, of course, this entire line of reasoning can be challenged. Other philosophers have avoided the need for a Creator by supposing that a huge multitude of universes exists. Most of them would be barren, but somewhere among them would be at least one with the necessary properties for life, and this, of course, is the one that we live in. The Sour Lemon philosophy has been resurrected on the grandest possible scale.

But one escalation invites another. Lee Smolin has speculated that new universes are generated within black holes in existing ones. A type of natural selection among universes then operates to favor the production of those that can generate life. By this treatment, complexification has also been extended to the multiuniverse level, reentering the contest with the Creation and the Sour Lemon schools.

I would prefer to defer this argument until the time when we can detect these other universes, if they exist. By that time our descendants may have gained the wisdom needed to deal with them. As Martin Gardner has written, "I believe there are truths as far beyond our grasp as calculus is beyond the grasp of a cat." But we do have it within our ability to determine if a Life Principle exists within our own universe. We need to get out there and look.

Of course, the answer may be negative: We may examine a multitude of worlds and find no sign of self-organization, even in those where the necessary prerequisites appeared to exist. The Creation and Sour Lemon schools would then remain to contest the philosophical terrain. But perhaps the experience of exploring other worlds would give us the strength and sense of purpose to create a new dream—one suitable for a barren universe.

Before we celebrate or fall into despair, though, we have the business at hand. How can we best locate self-organization and life within our solar system? To start, we might want to reexamine what the solar system has sent to us.

7

Cosmic Sweepings

I watched my television screen vicariously as wave upon wave of incoming meteorites bombarded Dallas, ultimately leaving a huge crater in place of the downtown area. It was hard to take the destruction seriously, as it came in an era of Hollywood films when natural catastrophes are very much in fashion. I have seen dinosaurs prowling downtown San Diego, tornadoes of unprecedented magnitude demolishing the Midwest, and volcanoes erupting in the middle of Los Angeles. In the case of the meteorites, it was easy to dismiss such scenes as fantasy. To my knowledge, no human has been killed by falling debris from outer space. Yet silent evidence warns of the damage that can be done.

I have walked around a giant meteor crater more than a kilometer wide, in northern Arizona. This huge hole was caused by an object about 30 meters in diameter. In 1908, a meteor twice that size broke apart above an uninhabited area of Siberia, burning out a vast area of forest. Had it occurred over a city, buildings would have been flattened for many kilometers around. As I have mentioned, the extinction of the dinosaurs 65 million years ago was likely to have been triggered by an impact from an object ten times the size of the one that fell in Arizona. It landed in the Yucatan, in Mexico, leaving a crater 170 kilometers in diameter. The immediate results were a huge fireball, earthquakes, and tidal waves. The released dust then produced a months-long period of total darkness.

Writers Timothy Ferris and Paul Davies have provided vivid accounts of the consequences we might expect if such an impact happened now. In Ferris's narrative, a tidal wave half the height of the Empire State Building hits Manhattan, and other floods submerge most low-lying regions of the Earth. Debris tossed into space by the collision returns in a fiery display and ignites forests, villages, and cities. The following blackout kills most plant life and thus produces famine among the remaining human population. While some humans may survive, civilization will perish.

Films such as *Deep Impact* and *Armageddon* have also made the threat of extinction by heavenly objects vivid, and the asteroid 1997 XF11 will pass within 1 million kilometers of Earth in 2028, to remind us that the possibility of a new collision is real. In the last chapters, I will return to this topic to illustrate our need for long-range scientific planning with regard to our future. But for now, I want to reverse the image and show how space debris has been considered to be a possible messenger for life.

A Sign of Life

It was meteorites that were the key. The theory runs that they, or some of them, are the products of a smash up in space, the fragments of a lost planetary body. . . .

The meteorites, he says, keep coming in—big ones and little ones, raining in with the earth's attraction. Okay then, keep looking, keep watching. Some day you'll get an unconsumed mass of sedimentary rock off that vanished planet. Sedimentary rock, mind you, fossil-bearing rock. Get it and you've got the secret of the galaxies. One fossil speeding in from outer space, one bit of fossil life unknown to this planet, one skull from a meteor's heart and space out there . . . becomes alive. Man is no longer lonely. Life is no longer a unique and terrible accident. It, too, holds its place with the spinning suns.

The speaker is Jim Radnor, a sheepman-turned-interplanetary fossil hunter in Loren Eiseley's story "The Fifth Planet." Radnor subsequently called off his hunt. The destruction of Hiroshima and Nagasaki by atomic bombs left him disillusioned about the nature and direction of the Divine Plan. In real life, others have taken up this quest. Their expectations have been scaled down, however, as our understanding of meteorites has increased.

The most obvious nearby source of cosmic debris lies in the asteroid belt, a collection of hundreds of thousands of airless rocks that circles the

Sun between the orbits of Mars and Jupiter. Ceres, the largest asteroid, has a diameter of about 1,000 kilometers. Its surface area is about the size of India but, to most of us, much less interesting. Assorted bodies of various smaller sizes are represented in the belt, but if they were all lumped together, their combined mass would be only one-twentieth of 1 percent of Earth's. This amount of material will not serve to build a respectable planet, and the current opinion of astronomers is that the asteroids represent original solar system debris that never came together. Most asteroids occur in this belt, but some lie farther out and others (the ones that make us nervous) follow irregular paths around the inner planets, occasionally crossing the orbit of Earth.

Spacecraft fly-bys of the asteroids Gaspra, Ida, and Mathilde have shown them to be heavily cratered; they are the equivalent of the automobiles that we see in junkyards. The rubble from past collisions may account for most of the smallish bodies that hit our own atmosphere. When such bodies arrive, we call them meteors, or sometimes shooting stars, as the heat produced by their entry into our atmosphere makes them visible as bright streaks. I have developed a stiff neck on more than one occasion by trying to watch meteor showers on chilly summer evenings.

When meteors do not burn up fully, some solid fragments, which we call meteorites, reach the Earth. While the large colliders may cause a disaster, the small ones only cause controversy.

As a sample of the disputes that we are about to resurrect, I will present a remark attributed to Thomas Jefferson. When he heard of a description by a Yale professor of a meteorite fall in 1807, he commented, "I would rather believe that a Yankee professor would lie than stones fall from the sky." This skeptical tradition has continued, particularly with regard to any possible connection between meteorites and life. We will illustrate it with three choice specimens: the Orgueil and Murchison meteorites, and the recent case of ALH84001, the rock from Mars.

The French Connection

On the evening of May 14, 1864, the town of St. Clar in southern France was illuminated by a brilliant blue-white fireball. Loud cannonlike explosions were heard over the next several minutes, after the intruder had faded to a dull-red color. Over twenty black meteorite fragments, some weighing

Astronaut James B. Irwin salutes the U.S. flag at the Hadley-Apennine landing site of the *Apollo 15* mission. This 1971 photograph was taken by Astronaut David R. Scott. The contrast between the bleak, airless lunar surface and the products of human technology is striking. (NASA/Johnson Space Center).

Artist Don Davis has prepared this illustration of the Moon's inhabitants, as described in the "Moon Hoax" published by the *New York Sun* in 1835. Bat-winged men, unicorns, and a temple nestle within a protected valley set in the desolate lunar surface. The painting accompanied an article in *Sky and Telescope* magazine, October 1981. Copyright Don Davis.

New stars are being formed in these three pillars of hydrogen gas and dust that have been dubbed the Pillars of Creation. They are part of the Eagle Nebula, a larger cloud of gas and dust that is 7,000 light years distant. The central pillar is about three light years tall. Image: Jeff Hester and Paul Scowan (Arizona State University), NASA, and the (Hubble) Space Telescope Institute.

The Hubble Space Telescope produced this view, called the Hubble Deep Field, by examining a very limited speck of the sky over the course of ten days. A bewildering assortment of galaxies can be seen, in the form of spirals, ellipses, and other shapes. Some of them are observed as they appeared when the universe was relatively young. (NASA, Robert Williams, and the Hubble Deep Field Team, Space Telescope Institute).

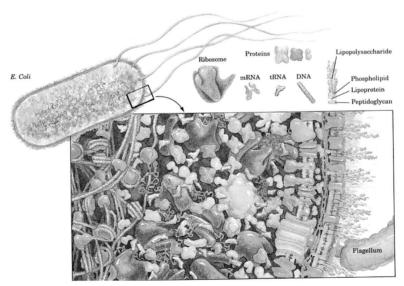

The upper-left corner of this diagram displays a cross section of a bacterial cell magnified about 50,000 times. In the remainder of the illustration, a corner of the cell has been increased in size by another factor of twenty, so that the total magnification is about one million. A bewildering tangle of tightly packed structures fills the interior of the bacterium. This artwork was taken from Donald Voet and Judith G. Voet, *Biochemistry*, Second Edition, 1995. Copyright John Wiley & Sons, Inc.

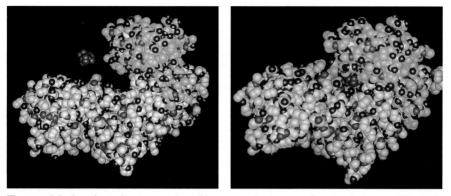

Two models that show the enzyme hexokinase enveloping a molecule of glucose. The balls of various colors represent atoms. On the left, the small violet glucose unit is approaching the "jaws" of hexokinase. On the right, the jaws have closed to engulf the glucose. This artwork was taken from Donald Voet and Judith G. Voet, *Biochemistry*, Second Edition, 1995. Copyright John Wiley & Sons, Inc.

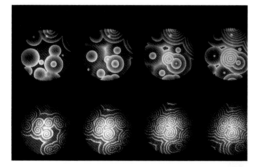

Complex spirals and whorls appear in simple chemical reactions when the appropriate ingredients are combined. These spatial patterns represent a simple form of chemical self-organization (or complexification). (Professor Arthur T. Winfree, University of Arizona).

A vision held by a number of scientists in the first half of the twentieth century about conditions on the surface of Venus is reflected in this painting by Ron Miller. The scientists felt that Venus was a warmer version of the Earth, with jungles, insects, mountains, swamps, and oceans under dense clouds of water. (copyright Ron Miller).

This 1980 painting by David Egge, *Venus Surface,* illustrates the altered perspective created by the Venera missions. A volcano, set in barren surroundings, contributes sulfur compounds to the planet's thick, acidic cloud layers. Artwork, David Egge, copyright 1980.

Voyager 2 obtained this picture of Jupiter on July 7, 1979. It displays a portion of the southern hemisphere, including the Great Red Spot. (Image: NASA/JPL/Caltech).

This photograph of Jupiter's moon Europa was taken by the Galileo spacecraft. An ocean may lie beneath Europa's fractured icy surface. (Image: NASA/JPL/Caltech).

The Galileo spacecraft photographed an area of Europa that displays features resembling the ice floes seen in the polar seas of Earth. The dark linear, curved, and wedge-shaped bands suggest a thin, icy crust with soft ice or liquid water beneath it. (Image: NASA/JPL/Caltech).

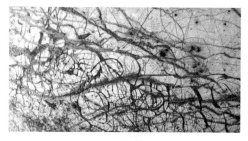

Artist David Egge has created this view of Europa's channeled, fractured terrain as it might appear from Europa's surface. Jupiter and Io can be seen in the black sky above. (Artwork, David Egge, copyright 1997).

In this painting by a NASA artist, a "hydrobot" (a self-propelled underwater vehicle) is exploring a hot vent at the bottom of Europa's hypothetical ocean. The hydrobot was deployed by a cryobot (at the upper left) that melted its way through the icy crust of the planet. (NASA/JPL/Caltech).

This view of Mars is centered on the huge Valles Marinesis canyon system, which extends for thousands of kilometers across the surface. It is a mosaic of 102 *Viking* orbiter images and presents Mars as it would be seen from a spacecraft 2,500 kilometers above the planet. (Astrogeology team, U.S. Geological survey, Flagstaff, Arizona).

The rover *Sojourner* sits on the *Pathfinder* lander, before the ramps were deployed to permit it to descend to the surface. The airbags used during the landing are visible, and the rocky, barren landscape of the Ares Valles region of the Mars mission furnishes the background. (JPL/NASA). (Sojourner, Mars Rover, and spacecraft design and images copyright (c) 1996–97, California Institute of Technology. All rights reserved. Further reproduction prohibited.)

Distant Shores, a painting by Pat Rawlings. Astronauts have landed in the Ganges Chiasma region of Mars and have set out in a rover to explore their environment. After driving a short distance, they stop to check out their systems. (NASA Artwork by Pat Rawlings/SAIC).

Enceladus, a moon of Saturn, has a relatively smooth surface and may, like Europa, hold an inner ocean. In this painting by William K. Hartmann, *On Enceladus, looking away from Saturn,* the Saturn moons (from left to right) Rhea, Tethys, Titan, and Dione populate the sky. This work was excerpted from the revised edition of *The Grand Tour: A Traveler's Guide to the Solar System,* by Ron Miller and William K. Hartmann. Used by permission of Workman Publishing Co., Inc. All rights reserved. Copyright Ron Miller and William K. Hartmann.

In this NASA painting by artist Craig Attebery, the *Huygens* probe descends by parachute onto the icy rocks and hydrocarbon-filled seas of Titan. The *Cassini* orbiter can be seen in the sky, as can Saturn and its rings. In reality, the haze will be too thick to permit Saturn to be visible by ordinary light. (NASA/JPL/Caltech).

Mimas, the innermost of the large moons of Saturn, contains an enormous crater that is about one-third its diameter. The view from this crater, named Hershel, may someday be one of the major tourist attractions of our solar system. In this painting of Hershel crater by a NASA artist, Saturn extends above the cliffs that ring the crater, and the *Cassini* spacecraft can be seen in the sky. (NASA/JPL/Caltech).

more than 2 kilograms, fell over an 18-kilometer path that included the villages of Orgueil and Nohic. One of them is said to have landed in an attic and burned the hand of the farmer who touched it. Others were collected by villagers in the fields. They were now available for the scrutiny of scientists, who spent the next century arguing about what they contained.

The classification of the meteorite, named for the village of Orgueil, was not a subject of dispute: It belonged to a small group that contained water and carbon compounds among their minerals. Such meteorites bore the formidable name of carbonaceous chondrites. The word *chondrite* referred to a feature shared by many meteorites: They held tiny rounded blue-gray mineral inclusions called chondrules. These inclusions and others appeared to have been heated strongly at some point in their history, unlike the dark, fine-grained, carbon-containing matrix that contained them. It is this matrix that concerns us here.

Many of the clays and other minerals in this group resembled earthly ones that were formed underwater. Many scientists concluded, by anology, that the meteorite minerals had also been deposited or altered underwater. Perhaps these alterations had taken place when they were part of a larger "parent" asteroid. Others disagreed, maintaining that the minerals had been formed directly from the solar nebula in their present state. In the case of the Allende meteorite, which fell in Mexico in 1969, the controversy was described as follows: "Thus, the essence of the debate is whether Allende has almost never seen a molecule of water in its history or whether it has, in fact, evolved in an environment something akin to the Florida Everglades, saturated with water."

This dispute over the history of such meteorites pales, however, before the one concerning their possible relevance to life. The Orgueil meteorite was the largest carbonaceous chondrite known until 1950, and scientists began their analysis of it immediately after it fell. A study published in that same year of 1864 declared that Orgueil contained materials "analogous to those of the organic part of several varieties of peats and lignites." These substances are produced by the decay of vegetal material on Earth.

Several years later, another analysis reported that Orgueil contained substances "comparable to the oils of petroleum." In the nineteenth century, such carbon-containing substances were believed to arise from living organisms, and these results were interpreted as evidence of extraterrestrial life. As we shall see, several prominent scientists at that time endorsed the idea that life on Earth had first been seeded by a meteorite.

Matters came to a climax in 1880 when Otto Hahn, a lawyer with some knowledge of geology, prepared thin slices of meteorites and examined them under a microscope. He claimed to see tiny fossils of corals and sponges. A German zoologist, Dr. D. F. Weinland, was sufficiently impressed to declare that in viewing Hahn's illustrations "we can actually see with our own eyes the remains of living beings from another celestial body." The meteorites that Hahn selected were ordinary chondrites, however, not those rich in carbon. His "discoveries" provoked a hail of refutations, as well as demonstrations that the structures that he had observed were, in fact, crystalline minerals. Hahn's conclusions were described as "so fanciful and so obviously without foundation in fact," and he was portrayed as an individual whose "imagination has run wild with him."

After this episode, claims concerning the remains of life in meteorites subsided, but they reemerged in the 1960s, when the growing importance of our space program stimulated a new interest in extraterrestrial objects. Analytical techniques had also improved immensely in the interval.

At a March 1961 meeting of the New York Academy of Sciences, Bartholomew Nagy (a Fordham University chemist) and his colleagues claimed that Orgueil contained chemicals typical of petroleum, which they regarded as a signature of life. They drew the inference that "biogenic processes occur and that living things exist in regions of the universe beyond the earth."

In further publications, they extended their analyses to the description of biological forms within the meteorite, citing "organized elements" that resembled "fossil algae." A sketch that accompanied a publication in *Nature* appeared to me as a ring of pebbles surrounding a hexagon, with amoeba-like forms inside and two Christmas-tree-type light bulbs on the outside.

Once again, such claims met a vigorous counterattack. One problem came from the obvious contamination of the meteorite specimen by Earth biology. A generation later, chemists using even more refined instruments could report that Orgueil "shows traces of what are believed to be indigenous amino acids . . . underlying what is probably extensive biological contamination. This finding is not surprising given the long terrestrial residence of this meteorite and the decrepit condition of the specimen analyzed." Somewhat earlier, a colleague at a meeting had passed this question to me: "When a meteorite falls in a sheep field, what do you expect to find on it?"

The forms that Nagy and his colleagues had observed received no more appreciation than the chemical contaminants that they reported. A group from the University of Chicago published its rebuttal in *Nature:*

> Many paradoxes are encountered if the organized elements are assumed to be of biological origin. The strongest argument in favor of such an origin is the morphology of the organized elements. But arguments based on morphology involve a great deal of subjective judgement. . . . Meteorites have long been notorious for containing structures resembling fossils, and although the work of Claus and Nagy was done with much greater care and competence than the early work of Hahn, the decision whether a certain form is of biological or inorganic origin is once again quite subjective. A purely inorganic mode of origin seems perfectly possible.

These elements that Nagy and his co-workers had described did not impress this group: "These spherical particles showed certain properties that were not entirely consistent with their identification as the remains of living organisms . . . [and] from these observations we conclude that these two types of particles are nothing but supercooled liquid droplets of sulphur and hydrocarbons."

Such problems have not been limited to extraterrestrial materials. Efforts to identify ancient microfossils on Earth have also been fraught with difficulties, even when the specimens could be placed within a known geological setting. Most of the published claims for supposed microfossils that were more than 2.5 billion years old were subsequently rejected.

To gain some idea of the difficulty involved, we can examine a comprehensive review by J. William Schopf and Malcolm R. Walter. (The first author is a UCLA geologist who is perhaps the most respected authority in this field.) They examined the claims put forward for forty-three specimens that were over 2.5 billion years old and concluded that "only two are considered to include unequivocal microfossils."

The winners were two collections from Australia. The remainder were classified as "dubiomicrofossils" and "nonfossils." Those considered nonfossils originally had been described as "yeast-like microfossils," "blue-green alga-like filaments," "spore-like microfossils," "filamentous bacteria or fungi," and "seven species of fossil protozoans." By contrast, the classifications provided by Schopf and Walter for these materials included "graphite particles," "bubbles and similar artifacts and modern bacterial contamination," "artifacts of preparation," "filamentous mineral crystallites," "mineralic pseudofossils," and "modern contaminants and artifacts of weathering and preparation." Schopf had been a coauthor of some of the original claims.

Doubt remained even about those two groups that passed inspection. Recently, however, Schopf has documented an unequivocal microfossil group about 3.5 billion years old from a site in Western Australia. Eleven separate species were documented, many of which resemble modern blue-green algae in their shapes.

Now that we understand how difficult microfossil identification can be with samples that originated on Earth, we can return to the Orgueil meteorite. After many claims and counterclaims had been made, Nagy sensed that he was losing the struggle. In 1975 he wrote that a biological interpretation of his organized elements in meteorites was only "a remote possibility." He noted that "there seems to be a growing tendency to consider them indigenous, non-biological organic particles which may be associated with mineral matter." Fortunately, by that time, nature had provided scientists with fresh, uncontaminated material with which they could continue the struggle.

Fire over Australia

On Sunday morning, September 28, 1969, a brilliant fireball appeared in the skies over Victoria, Australia. To a television technician, it appeared bright orange, with a silvery rim and a dull orange conical tail. A loud thunderlike explosion followed, and fragments rained down over a 16-kilometer-long ellipse near the town of Murchison. Over 80 kilograms (180 pounds) of material were collected. A number of them were taken soon after the impact, which minimized the chance for biological and chemical contamination. The meteorite proved to be a rich source of organic compounds, and extensive analyses followed, with varying interpretations.

Those inclined to seek signs of life could take comfort in the fact that amino acids were well represented. We can recall that a particular set of twenty of these substances is used to construct proteins by all organisms on Earth. In the meteorite, however, seventy-four different kinds were found in the initial analyses. This group included eight of the set of twenty used in protein, and eleven that are less common in our cells, but also fifty-five that have no role in our biology. The presence of this last group (together with the measurement of carbon, hydrogen, and nitrogen isotope ratios in the chemicals) was part of the proof that the results were not due to biological contamination.

The amino acids were not the major product, as they had been in the famous simulation by Stanley Miller and Harold Urey. They were accompanied by hosts of other types of organic substances (but no nucleotides— the raw material needed for RNA world). Large amounts of insoluble tarlike materials were present as well. A detailed review described the results as "complete structural diversity" and "synthesis by a random combination of C (carbon) atoms." Analyses of other meteorites of this type have provided results that differ in their details but not in the overall picture that they present.

What the meteorites have given us is a benchmark for unorganized material, an example of what carbon chemistry will provide when no forces that produce self-organization have been at work.

At a point when it appeared that further analysis of the Murchison meteorite might be left to those who enjoyed compiling exhaustive catalogs, a sudden surprise emerged. As I mentioned earlier, one prominent property of the amino acids and most other building materials used by life on Earth is their handedness (the words *chirality* or *optical isomerism* are often used in scientific papers). I mean by this that they can occur in two forms that are mirror images of one another but cannot be superimposed upon each other, as is the case for your right and left hands.

In constructing proteins, our cells reach exclusively for the left-handed amino acids and shun the right-handed ones. When amino acids are constructed from simpler materials in the laboratory, however, both left- and right-handed forms are produced in equal amounts. The early analyses of the Murchison meteorite also appeared to provide equal amounts of both mirror-image forms. In the early 1980s, however, Bartholomew Nagy (then at the University of Arizona) with a new colleague, Michael Engel, reported that some of the Murchison amino acids used in Earth biology occurred with an excess of the left-handed form. They did not attempt to interpret this as a sign of extraterrestrial life, but just the same, others claimed that this was the result of contamination by materials from Earth.

Engel continued the work over the next decade and a half, however, and by the end of that time he had apparently demonstrated by isotope studies that his results were real (though some possibilities for error still remained open). At about the same time John Cronin's group, one quite skilled in meteorite analyses, obtained similar results for nonbiological amino acids. Thus handedness was found in amino acids that are products of our biology and some others that have nothing to do with life on Earth. Some unknown process was at work.

One spectacular suggestion was put forward: Our solar system passed near a neutron star in its infancy, and this event biased the synthesis of the chemicals within it. If this is correct, then we might expect left-handed amino acids to predominate everywhere in our solar system, but not necessarily in others (it will take awhile to check this out). Other scientists have argued that the neutron star is not needed, and that scattering of radiation by dust particles during the formation of our solar system would produce the same result. Whatever the cause may be, we have learned an important lesson from the Murchison meteorite. It has taught us nothing about self-organization, but has warned us that a property we regard as a sign of life may result from other causes.

This victory on one claim concerning meteorites has served to renew others. At a 1997 San Diego meeting that overflowed with enthusiasm over the results from a Martian rock (see the following section), an organizer reported that Murchison contained "mushroom-shaped" structures with "stalks" and spheres that resembled spores. Shall we conclude that life has somehow infected the rubble of the asteroid belt? The carbon chemistry of the meteorite has sent us just the opposite message. We must fall back on one of our rules: Extraordinary claims require extraordinary proof.

We have seen how easily claims based on microscopic shapes can go wrong, even for microfossils from this planet. For a claim of extraterrestrial life based only on form, we would want an authentic "smoking gun" such as a skull or clear imprint of a detailed invertebrate, especially one unknown on Earth. There has been no stampede of scientists eager to endorse claims of life in Murchison. Yet meteorite studies took on a new vigor in 1996, when NASA called a press conference to present several lines of evidence that were present in a specimen found in the Antarctic.

The Rock from Mars

The Orgueil and Murchison meteorites had arrived with the flash of fireballs and the boom of thunder, startling the inhabitants of the areas. Many others may have fallen without as great a display, or with one that went unnoticed for lack of observers. In such cases, the fragments may simply have taken their places among their Earth-bred brethren without attracting further notice. But certain meteorites have chosen a select landing area that has given them public attention many centuries after their descent: the ice-covered plains of Antarctica.

Every year since 1976 a small group of volunteers has traveled south to McMurdo Sound base during the Antarctic summer. They participate in the Antarctic Search for Meteorites (or ANSMET) Project conducted by the National Science Foundation, NASA, and the Smithsonian Institute. They are flown with their camping equipment and vehicles into the interior, where they drive their snowmobiles in formation along the ice sheets, searching for the readily visible dark rocks.

The specimens that they hunted arrived much earlier. They fell thousands of years ago and were buried in compressed snow. In time the snow became ice, which turned bluish in color as it aged. Gravity moved the rocks several meters each year from the interior toward the coast. If their journey was blocked by intervening mountains, they were thrust up and eventually exposed at the surface, after the ice above them had sublimed away. The 17,000 meteorites collected in this way outnumber by at least sevenfold those retrieved after an observed fall.

A particular meteorite of this type was collected by Dr. Roberta Score in 1984 in the Allan Hills area, not far from McMurdo Sound, "while four other fellow researchers were taking a break and having some fun riding on their snowmobiles." Pale greenish-gray in color and potato sized, the 2-kilogram rock was named ALH84001, after its site and year of collection. It was packed into a sterile plastic bag for eventual shipment, along with 100 companions, to the Johnson Space Center in Houston. No special attention was paid to it at the time, for it was thought to come from the asteroid belt. A dozen years later, I was able to observe a fragment of it in a glass case kept under armed guard, at a symposium called in its honor at George Washington University. Obviously, some significant events had taken place in the interval.

The first honor was delayed until 1994 when it was classified among the SNC meteorite group, objects believed to have originated on Mars. This group gets its name from the sites Shergotty-Nakhla-Chassigny, locations in India, Egypt, and France, where the first meteorites of this type were collected between 1815 and 1911. The initial set of about a dozen samples was grouped together first on the basis of their similar mineral composition and later on their isotope ratios. When one of them (EETA79001) was later found to hold trapped gas bubbles whose composition resembled the Martian atmosphere (as measured by the Viking landers), it was decided that the entire group had originated on Mars. ALH84001, with its age of 4.5 billion years, was much more ancient than its fellows, which originated from 180 million to 1.3 billion years ago. Its membership was based on mineral analogy rather than trapped gases.

With its origin established, the various radioactive isotopes within ALH84001 could provide a limited biography for it. It crystallized from a melt as part of a larger rock formation shortly after Mars was formed. Between 3.9 and 3.6 billion years ago, the rock formation was fractured by a nearby impact. After that shock water percolated through the cracks in the rock, and at a later time, globules of a carbonate mineral formed in those locations. We know this mineral as limestone; it can form when excess carbon dioxide is present in water. Further impacts, between 1.3 billion and 16 million years ago, fractured the globules. Finally, about 16 million years before the present, another hit by a meteorite dislodged ALH84001 from its rock bed and sent it hurtling into space. After wandering for millions of years, it landed in Antarctica 13,000 years ago, to await its very recent discovery.

This modest rock rose to celebrity status on August 7, 1996. A press release on that day announced that "a NASA research team of scientists at the Johnson Space Center and at Stanford University has found evidence that strongly suggests primitive life may have existed on Mars more than 3.6 billion years ago." On the same day President Clinton held a press conference on the South Lawn of the White House, following another held by NASA scientists. The President commented that "today rock 84001 speaks to us across all those billions of years and millions of miles. It speaks of the possibility of life. If this discovery is confirmed, it will surely be one of the most stunning insights into our universe that science has ever uncovered. Its implications are as far-reaching and awe-inspiring as can be imagined." The President called for a space summit to be held "to discuss how America should pursue answers to the scientific questions raised by this finding."

An editorial in *Nature* the next week captured the spirit of the event: "The space science community has just enjoyed one of its finest weeks imaginable, basking in the reflected glory of a result that tentatively suggests that there was once life on Mars—capped by a ringing endorsement of its work delivered by an ebullient US president on the White House lawn." On August 16, the technical details were published in *Science*. What had been found?

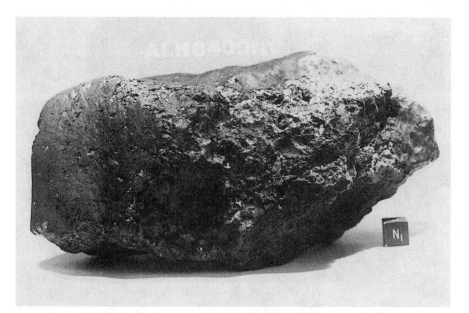

This meteorite, ALH84001, was discovered in the Allan Hills area of Antarctica in 1984. It is believed to have been dislodged from the surface of Mars by a huge impact, and to have fallen on Earth 13,000 years ago. (NASA/Johnson Space Center)

No skull or rich harvest of amino acids was detected in the rock. The bulk of evidence came from the orange-brown carbonate globules that lined the cracks in the rock and tiny grains of magnetic iron oxides and iron sulfides that nestled nearby. The patterns of the linked minerals resembled those produced by microorganisms on Earth.

Also present were representatives of a group of organic compounds called PAHs (for polycyclic aromatic hydrocarbons), which occur in coal and petroleum. The type and abundance of the specific organic molecules were "suggestive of life processes." Finally, segmented wormlike objects that resemble fossilized bacteria were observed using a high-resolution electron microscope. The scientists' conclusion was that these shapes were the remains of the organisms that had deposited the minerals and decomposed to form the chemicals.

In concluding the *Science* paper, the authors remarked, "None of these observations is in itself conclusive evidence of past life. Although there are alternative explanations for each of these phenomena taken individually, when they are considered collectively, particularly in view of their spatial association, we conclude that they are evidence for primitive life on early Mars."

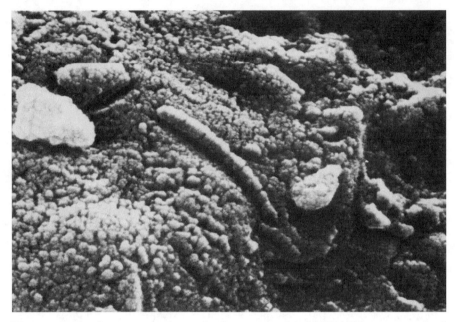

These tubelike forms were observed when meteorite ALH84001 was examined with a high-resolution scanning electron microscope. Scientists have debated whether the forms represent microfossils of early Martian life or mineral formations unrelated to life. (NASA/Johnson Space Center)

If you have been following the thrust of this chapter, you might anticipate that others would arrive at a different conclusion.

The next sentence from the *Nature* editorial cited previously remarked that "parts of the community respond as though they had just been mugged." They certainly did not take the events quietly. In the following months, we heard that the carbonates may have been formed at high temperatures, that inorganic processes can easily account for the other minerals, that the microfossils are really "whisker-shaped" geological crystals, that the shapes of such structures are at any event very weak evidence for past life, and that the PAHs may be contaminants that seeped in from the Antarctic ice. Such compounds are widely distributed as pollutants on Earth; they also form spontaneously in interstellar space.

Perhaps the weakest counterargument was that the fossil bacteria were too small, tinier than their earthly counterparts. Tiny Earth bacteria were then promptly located in rebuttal. In addition, it is hard to predict what an appropriate size should be for Martian organisms of unknown composition.

We will capture the flavor of the debate with a few samples. For example, meteorologist Edward Anders of the University of Chicago had been a critic of the Orgueil claims of the 1960s. Among his comments on ALH84001 were the following:

> The Orgueil meteorite was bad data and bad interpretation. Now we have good data and bad interpretation. . . .
>
> This is half-baked work that should not have been published. . . . For all these observations, an inorganic explanation is at least equally plausible, and, by Occam's razor, preferable. Consistency arguments alone—weak consistency arguments especially—cannot strengthen, let alone prove, an extraordinary conclusion.

Derek Sears, editor of *Meteorics and Planetary Science,* added his opinion: "These arguments are flaky and simplistic. Weathering is a sloppy process. Things leach in, then leach out. They do not do the obvious. . . . There are nonbiological explanations of McKay's data that are much more likely." John Kerridge, a University of California at San Diego planetary scientist, added, "These minerals formed at such a high temperature that life could not have survived—on Earth or on Mars."

Jack S. Farmer, a NASA investigator involved with the search for life on Mars, also joined the fray: "The problem is that at that scale of just tens of nanometers, minerals can grow into shapes that are virtually impossible to distinguish from nanofossils."

Finally, we will turn to Everett Shock of Washington University: "And look at the Murchison meteorite, thought to come from the asteroid belt. Hundreds of organic compounds have been identified in it, including amino acids and compounds closer to the things organisms actually use. It has carbonate minerals in it, too—and real solid evidence of water—yet there isn't anybody saying that there is life in the asteroid belt."

These last comments were made in 1996, however. Had Dr. Shock waited until the summer of 1997, he would have encountered exactly that claim despite the absence of the extraordinary proof needed to sustain it.

Life after ALH84001

Neither side in the debate over the significance of this meteorite has yielded appreciable ground, and the dispute promises to continue in the journals indefinitely. Little has been heard of government-sponsored space summits or new initiatives. NASA's budget has been threatened by further cuts but appears to have stabilized at this writing. In late 1996, three Martian meteorite fragments were offered at auction, but the bids failed to reach the required threshold of $1.5 million for the collection, and no sale was made. A *New York Times* advertisement of November 11, 1997, announced that one-tenth-carat samples of the Zagami Martian meteorite were available for $98 (plus shipping and handling). I do not know how sales went, but the market does seem to be deflating. What can we conclude from these events?

We have seen that the question of extraterrestrial life lies at the heart of a tug-of-war between three different visions of our place in the universe. The way that we behave in the future may be affected at a deep level by our perception that the cosmos is fertile or barren. But questions of this magnitude cannot be settled by weak evidence.

Meteorites so far have not provided us with the necessary beef. We may grow lucky in the future and come up with a viable spore or an extraterrestrial body part, but I suspect that they will not be had so easily on this planet. In fact, meteorites may make it harder for us to answer the deepest questions concerning life.

We have learned one important lesson from the Martian meteorite collection. Rocks can be blasted off that planet by a well-aimed collision, and make their way to Earth with their interiors relatively intact. (Another set of Antarctic meteorites came to Earth from the Moon.) Logically the process could also be reversed. Though Earth has higher gravity than Mars, some fragments might occasionally be liberated from our planet by collisions and find their way to nearby worlds. An interplanetary rock shuttle appears to exist for the inner solar system, but with an unknown schedule.

The shortest transfer time recorded for a Martian meteorite has been 700,000 years. But in a computer simulation by Joseph Burns of Cornell University and his colleagues, times of 16,000 years appeared reasonable. If everything fell right in their simulation, the journey could be shortened to just a few years, but the odds against this were estimated as 10 million to 1.

If the trip were short enough, living microbes or their surviving spores might be part of the delivery. Their chances would depend on how well they were protected from cosmic and ultraviolet rays inside the rock. One piece of data is revealing. The *Apollo 12* astronauts retrieved a television camera from a nearby *Surveyor III* spacecraft and returned it to Earth. Freeze-dried microorganisms were then discovered that had survived for

950 days on the Moon within the camera's polyurethane insulation. When thawed back on Earth, they reproduced and lived normally.

Presumably, the same bugs could have survived a meteorite trip of the same duration. As we shall see, some theories hold that Mars and Earth had similar climates about 4 billion years ago. Early in the history of our solar system, then, our earthly biosphere may have "infected" Mars, or vice versa. As H. J. Melosh wrote in *Nature*, "Planets of the Solar System should therefore not be thought of as biologically isolated: from time to time large impacts may inoculate Mars and the other planets of the inner Solar System with a sample of terrestrial life." I will call this possibility local panspermia, to distinguish it from a more widespread possibility that we will consider shortly.

The possibility of local panspermia confounds the search for the Life Principle by providing an alternative route for the origin of life on Mars. In the argument over ALH84001, both sides sometimes assumed that panspermia was at work. The mineral evidence was interpreted with reference to known terrestrial bacteria, and the arguments that concerned the small size of the fossil microbes also used our own bacteria as reference points.

Stanford Professor Richard Zare of the ALH84001 team brought up the larger question: "If we are right and life evolved on Mars during an early period in its history when it had an atmosphere and a climate similar to Earth's, then it is likely that life evolved on the countless other planets that scientists now believe exist in our galaxy." But to arrive at Zare's conclusion, we must discover organisms on Mars or elsewhere that have a separate origin. Local panspermia would extend the range of our biosphere but not tell us about the universe.

The Roots of Panspermia

The idea of panspermia dates back more than a century, and its early eloquent advocates did not intend to limit it to nearby worlds. I will quote from an address delivered in Edinburgh, Scotland, in 1871, at a time when belief in spontaneous generation had been abandoned by all but a few entrenched defenders.

> Careful enough scrutiny has, in every case up to the present day, discovered life as antecedent to life. . . . Dead matter cannot become

living without coming under the influence of matter previously alive. This seems to me as sure a teaching of science as the law of gravitation. . . . I am ready to adopt, as an article of scientific faith, through all space and through all time, that life proceeds from life, and from nothing but life.

How, then, did life originate on the Earth? Tracing the physical history of the Earth backwards . . . we are brought to a red-hot melted globe on which no life could exist. Hence when the Earth was first fit for life, there was no living thing on it. There were rocks solid and disintegrated, water, air all round, warmed and illuminated by a brilliant Sun, ready to become a garden. . . .

When a volcanic island springs up from the sea, and after a few years is covered with vegetation, we do not hesitate to assume that seed has been wafted to it through the air, or floated to it on rafts. Is it not possible, and if possible is it not probable, that the beginning of vegetable life on Earth is to be similarly explained? Every year thousands, probably millions, of fragments of solid matter fall upon the Earth—whence came these fragments? . . .

It is often assumed that all, and it is certain that some meteoric stones are fragments which had been broken off from greater masses and launched free into space. . . . When two great masses come into collision in space it is certain that a large part of each is melted; but it seems also quite certain that in many cases a large quantity of *débris* must be shot forth in all directions, much of which may have experienced no greater violence than individual pieces of rock experience in a land-slip, or in blasting by gunpowder. . . . Hence, and because we all confidently believe that there are at present, and have been from time immemorial, many worlds of life besides our own, we must regard it as probable in the highest degree that there are countless seed-bearing meteoric stones moving through space. If . . . no life existed upon this Earth, one such stone falling upon it might, by what we call natural causes, lead to its becoming covered with vegetation.

In this speculation, the origin of the "seeds" is pushed back before the start of our solar system, into the greater universe. I will call this concept cosmic panspermia, to distinguish it from the more limited variety. This removal does not solve the question of life's initial origin, of course. The speaker also provided his view on that matter: "Overwhelmingly strong proofs of intelligent and benevolent design lie all around us, . . . showing to us through nature the influence of a free will, and teaching us that all living beings depend on one ever-acting Creator and Ruler."

The meeting at which these words were spoken was not a science fiction convention or an assemblage of theologians. The British Association for the Advancement of Science was listening to the Inaugural Address of its President, Sir William Thomson (1824–1907), who later became Lord Kelvin. He was one of the foremost British physicists of his century, with contributions in the areas of heat, magnetism, and electricity, among others. His remains are preserved in Westminster Abbey and his name in the absolute scale of temperature used by physicists.

Thomson was not alone, or even necessarily first in this idea. In the same year the accomplished German physicist Hermann von Helmholtz (1821–1894) voiced similar thoughts:

Who can say whether the comets and meteors . . . may not scatter germs of life wherever a new world has reached the stage in which it is a suitable dwelling place for organic beings? We might, perhaps, consider such life to be allied to ours, at least in germ, however different the form it might assume in adapting itself to its new dwelling place.

Despite the reputation of its advocates, the idea did not sail smoothly across. Shortly after Thomson's address, a colleague of Charles Darwin, Joseph Dalton Hooker, wrote to Darwin:

The notion of introducing life by Meteors is astounding and very unphilosophical. . . . Does he [Thomson] suppose that God's breathing on Meteors or their progenitors is more philosophical than breathing on the face of the earth? I thought that Meteors arrived on the earth in a state of incandescence. . . . For my part I would as soon believe in the Phoenix as in the Meteoric import of life.

We now appreciate that the interiors of meteorites may remain quite cool, preserving their contents, even though the surface is heated strongly.

Other critics suggested that a meteorite was too heavy to be expelled from a large planet. In the first decade of this century, Svante Arrhenius (1859–1927), a Nobel Prize winner in chemistry, tried to avoid these problems by eliminating the vehicle. He proposed that a lightly shielded microorganism could be ejected from a planetary atmosphere by lightning and be propelled out of its solar system by the pressure of radiation from its star. It would survive in space as a dormant spore.

New questions arose, of course. How long could a spore survive exposure to ultraviolet light and cosmic radiation? Carl Sagan and others have argued against the possibility of such an event. I reconsidered the odds, however, when I read of a strain of Earth bacteria that can survive 3,000 times the human lethal dose of ionizing radiation.

Although their chromosomes were shattered by the exposure, the microbes kept them in multiple copies and used clever repair mechanisms to stitch the pieces together. Their preferred living quarters were also remarkable. The creatures had been isolated from the "feces of some elephants and South American llamas, in some samples of Swedish underwear, in Antarctic rocks and in water tanks used as shielding against lethal radiation from pieces of cobalt-60." After such a training schedule, a trip to Alpha Centauri would seem a picnic excursion for these bugs. New calculations have, in fact, suggested that bacteria encapsulated in a light dust jacket might survive a million-year trip to a nearby star system.

So proponents of Arrhenius's scheme still advocate it, although as we have seen, the case for meteorites as shuttles has also been restored by the safe arrival on Earth of ALH84001. Panspermians can select the vehicle they want as a matter of taste, since science has little to add at the present moment. Suppose they should be right? How would this impact on our previous contest of Creation versus Cosmic Evolution versus the Sour Lemon School concerning the place of life in the cosmos?

Lord Kelvin illustrated that panspermia is compatible with a belief in Creation. The idea that seeds of life have drifted from star system to star system says nothing about their ultimate origin—this still could be left to God. This vision is perhaps closest in spirit to that of the sixteenth to nineteenth centuries, which pictured a universe that was filled with Earth-like worlds and populated by Earth-like creatures. Such creatures made up a series that extended from the lowest forms to the Highest One as part of the Great Chain of Being. This resurrected Chain of Being would now be limited to a small fraction of the cosmos, but within that realm the eighteenth-century vision (and that of modern science fiction movies) could live again.

Cosmic panspermia would be a mixed blessing for the Sour Lemon School. On the one hand, it would offer improved odds for their long shot. Most geologists now believe that life started quite quickly on this planet—

almost immediately after the bombardment ended. If our origin represented a rare, lucky event, then the idea that it happened so quickly here, while not impossible, becomes intellectually unattractive. It is far easier to hold the view that trials were run for billions of years on countless worlds until the fortunate break took place. Once it had started, life then spread and took root here as soon as our planet was ready to receive it.

On the other hand, the vision of life as a hardy, migratory phenomenon weakens the basic emotional message of the Sour Lemon School. In their writings members of this school often picture life not only as something that was unlikely to have started in this hostile universe, but also as an aberration that is equally unlikely to persist.

The basic assumption of the Life Principle would not be weakened if panspermia exists. A universe that favors the spontaneous development of life could also be one that supports its propagation. But panspermia may confound efforts to show that life elsewhere has a separate origin. To settle the point, we may want to explore environments that are incompatible with our kind of life, to see whether other kinds of life have found them exactly to their taste.

The discovery of panspermia, whether cosmic or local, would not then settle the basic debate or provide the answer to life's origin, though it would add a new dimension to our explorations. Fortunately, we will not have to wait forever for the answer to this particular question, because it awaits us out in the solar system. Our search will also provide a critical test for another cosmology that, though based on panspermia, escalates the idea to create a unique and peculiar vision of its own. The most favored transportation vehicle in this novel idea system is yet another heavenly body: the comet.

Bright Stranger in the Night

I stood on my front lawn and looked up at the evening sky in the appropriate direction, but without much hope. I remembered how I had peered years earlier, hoping to get a glance first at Comet Kehoutek, and then at the celebrated Halley's Comet. I had seen nothing on either occasion. This time I was delighted to find, with little effort, a luminous smudge in the sky. With ordinary binoculars, I could easily make out a shining head and long forked tail. Comets existed and were not just a media hoax. For several weeks thereafter, I looked into the sky almost every evening, wherever I was, to find it. If the occasion arose, I would point it out to a companion or passerby. When Comet Hale-Bopp finally disappeared, I felt a vague sense of loss.

Others suffered a more severe loss. On April 28, 1997, the newspapers told of the thirty-nine members of the Heaven's Gate cult, who had jointly committed suicide in a suburb of San Diego. They took the appearance of Hale-Bopp as a sign that it was time for them to "shed their containers" and leave this planet. Suicide was a prerequisite for their rendezvous with an alien spaceship that was tailing the comet. This ship would take them home to a higher level of existence. The spectacular appearance of the comet had evoked a spectacular response.

Comets, like asteroids, coalesced out of the original material of the solar nebula, at the birth of our solar system. Their birthplace lay farther out, however, beyond the orbits of the planets, where cooler temperatures allowed them to keep their water, organic molecules, and other light materials. Hundreds of billions of comets may reside in the Oort Cloud, a vast collection that extends one-fifth the distance to the nearest star. Others originated in the Kuiper Belt, a collection of small, cold planetesimals in the region beyond the orbits of Neptune and Pluto.

Most comets stay in their home areas, but a few are displaced into elliptical orbits that bring them closer to the Sun. As they approach, their lighter materials evaporate into space, forming a tail that may extend for many millions of kilometers. At their heart, however, is a relatively small core of rock and ice.

Comets have often been regarded as messengers of important events or disasters. Shakespeare wrote in *Julius Caesar* (Act II, Scene 2), "When beggars die there are no comets seen; the heavens themselves blaze forth the death of princes." But comets have the power to cause disaster as well as advertise it. Collisions with comets, in addition to those with asteroids, may have caused some of the large mass extinctions observed on this planet. Some astronomers, understandably, have wanted to find a more attractive role for such beautiful objects. One fashionable idea uses comets as transport vehicles to bring water and organic materials to this planet before life began. This idea is opposed by other scientists, who feel that these substances were already present in our planet when it first came together.

However this controversy may resolve itself, it tells us little about the origin of life. The crucial unresolved question concerns how these materials got organized, rather than how they arrived. A few observations have been made on the contents of comets: They contain mixtures of small molecules and perhaps some larger ones of random, chaotic composition. No evidence for the Life Principle is present in the data. This circumstance has not inhibited Sir Fred Hoyle and his colleague Professor Chandra Wickramasinghe, who use the comets as starring players in their unique cosmology.

The Cosmos, According to Hoyle

In the beginning, there was the universe. It has always existed, with no starting point in sight. The Big Bang theory has it all wrong. Sir Fred considers that idea to be "like medieval theology." He even coined the name "Big Bang" to ridicule it, but the phrase caught on instead. For a discussion of its flaws, we can consult his autobiography:

> Big-bang cosmology is a form of religious fundamentalism, as is the furor over black holes, and this is why these peculiar states of mind have flourished so strongly over the past quarter century. It is in the nature of fundamentalism that it should contain a powerful streak of irrationality and that it should not relate, in a verifiable, practical way, to the everyday world.

Currently, Sir Fred favors a "quasi-steady-state" cosmology, in which a number of smaller bangs in selected places create new matter and maintain the density of the cosmos as it expands. A large outburst of this type, some 14 billion years ago, may have been confused with the creation of the entire universe, but that entity is much older. According to Sir Fred,

> The breathtaking complexity of life points to a universe of vast antiquity; it suggests that the universe has developed with respect to increasing order over a span of time that was enormous compared to the intervals usually contemplated in cosmology. . . .
>
> Because the universe has no beginning, according to this theory, since the history of the universe extends backward in time as far as one cares to go, the theory does not encourage the Judeo-Christian conception of a universe created by an external God. The steady-state conception is of a universe that contains within itself its own perfection of form, its own divinity, as one might say.

Sir Fred has concluded that the highest being, the one that controls all existence, is the universe itself. I will again allow his words to speak for him: "The overriding intelligence in the infinite future, which masterminds the development in our present time, must exercise its controlling influence simply to exist. . . . So it is with the universe, in which the controlling intelligence exists by virtue of the support the universe gives it."

These quotes may leave matters a bit vague, but in another place Hoyle is quite specific: "*God* is the universe: *God* ≡ universe."

This highest intelligence would, of course, control the lesser intelligences within it, in a relationship similar to the Great Chain of Being. At the next level below God, we would find an intelligence that determined the detailed laws of nature:

> Once we see, however, that the probability of life originating at random is so miniscule as to make the random concept absurd, it becomes sensible to think that the favourable properties of physics on which life depends are in every respect deliberate. The measure of intelligence needed to control the properties of the oxygen and carbon nuclei would be exceedingly high. It would be through exercising control over the coupling constants that an intelligence might determine a wide range of features of the universe.

Somewhat farther down the chain would be our own immediate Creator, an extremely complex and clever silicon chip: "It would not be possible for an intelligence, however great, to generate carbonaceous life without performing an immense amount of calculation. . . . The best way we know to perform the necessary calculations would be through the silicon chip."

But why would it choose to do this?

> We humans have now given rise in our turn to the silicon chip itself. So one would have the sequence (in which the arrows mean "leading to"):
>
> silicon chip → carbonaceous life → silicon chip
>
> by which the silicon chip would succeed in spreading itself.

This ambitious computer did not limit itself to a simple task, such as selecting a few suitable worlds and seeding them with life with a well-aimed vehicle. It chose instead to spread its creation, in the form of freeze-dried bacteria, helter-skelter throughout the galaxy. These bacteria, together with a liberal helping of the chemicals that were used to manufacture them, make up the interstellar dust clouds.

Most scientists prefer to interpret those clouds as mixtures of minerals, ice, tarry chaotic organic materials, and simple molecules that have virtually no relation to the set used by our own biochemistry. But Sir Fred is certain that they are wrong. He once commented,

> In my days as an active astronomer (by which I mean attending an intoxicating round of committee meetings in London) I used to remark that the community of astronomers lived in perpetual terror that one day it might stumble inadvertently on something important, a remark that did not greatly enhance my popularity.

To continue Sir Fred's narrative, the next developments occurred when a pregnant interstellar cloud gave birth to our Sun. The bacteria that were swept into our new Sun or found themselves too close to it were roasted. But others located safe havens at a greater distance from the inferno, within newly formed comets. Those comets with skewed orbits occasionally approached the Sun closely enough to melt their icy interiors, which created an opportunity for the bacteria to thaw and multiply. Providential encounters of comets with suitable planets then seeded these worlds with life. Thus began life on the Earth some 4 billion years ago. Bacteria also took up residence at the same time in the clouds of Venus, Jupiter, Saturn, Uranus, and Neptune, and they may also survive deep within glaciers on Mars.

It was not sufficient for the silicon chip beings, in seeking their purposes, to start life here. They further needed to ensure that evolution proceeded in the direction that they considered appropriate. To achieve this, they also scattered viruses carrying appropriate genes and other biological packages throughout space. When these packages arrived, they stimulated the next phase of evolution.

Some 570 million years ago, a very large comet delivered frozen eggs of multicelled organisms to us, triggering the Cambrian explosion. Subsequent parcels have brought in creatures as large as bees and other insects. Unfortunately, those species that reproduce using two sexes were excluded from this scheme. Hoyle felt it unlikely that both mates would be delivered by chance at the same time and place. This argument has come up before.

Lord Kelvin's 1871 comments to the British Association for the

Advancement of Science stimulated discussions that continued in later meetings. A report of the 1877 gathering included the following exchange: "Some one . . . introduced the Colorado Beetle, and this was held to be irresistibly funny. . . . At length another speaker arose to breathe the hope that when Papa Colorado Beetle dropped down on a meteorite he would leave Mama Colorado Beetle behind, which was felt far away to be the funniest thing of all."

Hoyle has argued that these random arrivals carry a different hazard: disease. Cancer results, for example, when a genetic message from space, meant to promote the budding of yeast, is accidentally absorbed by one of us. In Hoyle's words, "The phenomenon of cancer is an inevitable consequence of the present ideas." Other unhappy deliveries have wreaked havoc throughout human history. Hoyle's list includes periodic episodes of smallpox, a mysterious plague that ravaged classical Athens in 430 B.C., and the great influenza epidemic of 1918, among others. The AIDS virus, however, is specifically excluded. It "is such a strange virus I have to believe it's a laboratory product." Sir Fred feels that AIDS was first produced in a biological warfare program.

The preceding series of claims, if put forth by most of us, would end up in some editor's trash bin or see light in the pages of a lurid supermarket tabloid. Hoyle's narrative has received kinder treatment because of his notable earlier intellectual contributions in astronomy.

In the 1950s, he and several colleagues worked out the paths by which heavier elements are synthesized from lighter ones in stars. When one of the group, William Fowler, received the Nobel Prize in 1983, *Nature* magazine regretted that Hoyle had not been included in the award. In 1997, however, Hoyle did win the Crafoord Prize from the Swedish Academy of Sciences, which included half of $500,000. He has received many additional honors and awards, including his knighthood. He is the author of a very well-known astronomy text, many books of science for the public, and a number of science fiction novels. He served as the president of the Royal Astronomical Society and as vice president of the Royal Society, and he was a founder of the Institute of Astronomy in Cambridge, England. In 1992 a bronze statue of him was erected on the grounds of that institute.

Hoyle's achievements undoubtedly earned him a hearing for his theories, and his talents for communication ensured that his ideas were widely spread. As science writer John Horgan wrote in the *Scientific American*,

"Hoyle's personality no doubt adds to his appeal . . . he exudes a kind of blue-collar integrity. Hoyle strikes one as a man doggedly pursuing the truth, and to hell with the consequences. And he has a knack for sounding reasonable." Horgan opened his account with the comment, "In the dead of night, when the demons come, a special fear may creep into the hearts of scientists: what if Fred Hoyle is right? Then astronomy is a sham, biology a house of cards and modern medicine an illusion."

My own suspicion is that what many scientists fear in the dead of night is that they will never be able to free their disciplines of pseudoscience. Given the output from diverse sources such as astrologers, Creationists, and Sir Fred Hoyle, their own attempts to communicate to the public will be overwhelmed. A host of astronomers, disease epidemiologists, evolutionists, chemists, and others whose fields were reinterpreted by Hoyle's theories have issued detailed rebuttals, explaining at length in refereed journals why his proposals were not in accord with the facts, but this has not hindered the flow of books in which he repeats the same ideas.

I myself read in depth several of his papers that covered areas in which I had some experience: I concluded that the rule that extraordinary conclusions deserve extraordinary evidence had been totally trashed. Information-poor spectra that could be interpreted in a variety of far less controversial ways were cited as "irrefutable" evidence for their theories. When the intended fit of the data was less than perfect, Hoyle and Wickramasinghe did not hesitate to make adjustments to improve matters. Sometimes they fell into blatant errors in their handling and interpretation of tables.

In 1997, I met Wickramasinghe, Hoyle's coauthor on many of his books, at a conference. I brought up some examples of this type, but he was not perturbed. He simply acknowledged the mistake and asked why *I* was troubled by it. Later in the conference he went on to advocate some of the same ideas to a large audience.

Hoyle and Wickramasinghe shrug off past errors effortlessly. In fact, skepticism and self-doubt seem to play little role in their approach, even though they have reversed themselves completely on a number of topics. One source of their convictions was revealed in an interview of Hoyle by Brig Klyce, reported at a panspermia Web site on the Internet: "So I get all these [results] and now I'm unshiftable. I'm totally unshiftable now because it's sort of religion with me. That is the word of God. . . . It's there. It's like the road to Damascus, you know in the Christian Doctrine. Your eyes are opened. And I don't move from then onwards."

The religious certainty that pervades the views of Fred Hoyle is also reflected in his and Wickramasinghe's attitudes toward the views of others.

Doubtless there will be persons who never take a positive statement like this on trust, who would continue to argue, even as the snow closed over their heads, that an avalanche was not really bearing down upon them.

For when human beings refuse to distinguish propositions about the world that are demonstrably true from those that are manifestly false, we must surely be heading down a road to disaster.

Today we have the extremes of atheistic and fundamentalist views, and it is, in my opinion, a case of a plague on all their houses. The atheistic view that the Universe just happens to be here without purpose and yet with exquisite logical structure appears to me to be obtuse, whereas the perpetual quarrelings of fundamentalist groups is worse than that.

In this last statement we have arrived at an impasse common to religious quarrels throughout human history. Contending points of view entrench their own position, and no further progress is possible. We can escape from this tangle, however, by using the tool that first moved us away from the medieval cosmos: the scientific method. We can now move off our planet into the wider theater of our solar system and learn what nature may choose to teach us.

The theories of Fred Hoyle and the other less fanciful versions of cosmic panspermia have a common denominator: They predict that bacteria and viruses should be present throughout our solar system. If life has shifted readily from one star system to another in the past, then some living or dormant samples should be around at the present time. To capture them, we will need to collect interplanetary dust samples and probe within the interiors of likely meteorites and comets. As an alternative, we could search in the debris of impact sites on airless worlds, digging into shaded or buried regions that have not been exposed to solar radiation. Naked organisms might not survive a direct hit on an airless world, but some of the compounds from them might remain intact. No such material was encountered in crater debris samples returned during the Apollo Program, but a more extensive search could be made, when astronauts are next on the Moon again.

A Modest Agenda

Only a few explorations are now scheduled to examine this large question. The Near Earth Asteroid Rendezvous (NEAR) mission will approach the asteroid Eros on January 10, 1999 (one might have expected a Valentine's Day liaison, given NASA's record of landing Mars vehicles on Independence Day). Eros is more typical of the bodies that might deliver the kiss of death to us by collision, rather than of the carbon-rich asteroids that could give us clues about life. At any event, NEAR will do what its name implies, hover about and measure from a distance, rather than land and sample. The mission will end on February 6, 2000, with a slow crash landing on the asteroid's surface.

Of greater interest will be the Stardust mission, which will depart on February 15, 1999, fly within 100 kilometers of the nucleus of the comet Wild-2 in January 2004, and capture material shed from the comet. In addition, *Stardust* will pass through a recently discovered flux of fresh interstellar grains and sample this material as well. The spacecraft will make some measurements while en route, but most important of all, it will return the samples to us in January 2006. If the mission succeeds, we will gain a unique chance to let nature instruct us about the universe. If the cosmic panspermia idea is correct, then either or both samples should contain dormant microorganisms, or at least nonrandom collections of chemicals that bear a striking resemblance to our own biochemistry. If so, we will have to rethink our ideas about the universe in a dramatic manner.

A more likely result, in my opinion, is that both samples will contain the usual assortment that has been detected by indirect methods: some very simple organic substances, a mixture of slightly larger exotic chemicals bearing no relation to life on Earth, and some more complex tarry material. If this should be the case, then we will learn that neither the frigid depths of interstellar space nor the limited and usually frozen interiors of comets are suitable incubators for Earth life (or perhaps any other variety of life). In our search for the Life Principle, we will have to turn to the more promising theater of the planets.

8

A Plentitude of Worlds

> I dedicate this essay to an argument that evidence of
> life (albeit from my own domain of a paleontological past)
> may yet be found on another planet in our solar system—
> and that such a discovery might furnish the greatest
> plum of biological knowledge in all our history.
> —*Stephen J. Gould*

The ancients, using their unaided eyes, saw that seven bodies moved independently across the sky: the Sun, Mercury, Venus, the Moon, Mars, Jupiter, and Saturn. With the exception of the Sun and Moon, we call them planets, from a Greek term for wanderer. Their names could differ from one society to another, but their number was the same in all of them. The count remained fixed for thousands of years until January 1610, when Galileo turned his telescope on Jupiter and discovered four additional wanderers. They circled a planet rather than the Sun, yet three were larger than the Moon and one larger than Mercury. They were worlds in their own right. A human explorer on the surface of one of them would have support under his feet, a horizon in the distance, and the heavens above. I will use the term *planet* to apply to all of them, with only size as a limitation.

I have selected a diameter of 1,000 kilometers as an arbitrary cutoff for this group, and I will exclude the Sun, which is a star. Our collection holds a total of twenty-five bodies, if we include the Earth. Below the cutoff point, we enter the realm of the smaller moons and asteroids; they number in the tens of thousands. Our focus here will not be on their size or distance from the Sun, but rather on the bounty that they hold for human exploration and the search for life.

Why Go at All?

In the first chapter, I presented a quote from Kurt Vonnegut that held that only a "nightmare of meaninglessness" would be found beyond Earth. Historian Stephen J. Pyne has presented a deeper analysis of space travel as compared to past human exploration. He has considered the great age of discovery of Western civilization, in which the dimensions of our globe were revealed and separated human cultures became aware of one another's existence. The moral questions that arose from such meetings offered an opportunity for our civilization to examine its own values. The flood of new images and encounters that we harvested from geographical exploration gave us a chance to ask where we were headed in a human sense.

But when the oceans had been crossed, the habitable continents examined, and their people encountered, there remained only barren terrain, such as Antarctica, for further exploration. In Pyne's words, "The continent [Antarctica] was ringed by steepening gradients of energy and information, rich on the outside, barren toward the center; a journey to the source meant that there was more and more of less and less, until at the end there was only ice and self."

Our surveys of the solar system by robot spacecraft could be seen as an extension of the Antarctic experience, but one that revealed even more barren terrain: "But apart from the Earth, there was nothing. No life. No peoples. No intelligence." Pyne reasons, "With no distinctively human encounters possible, there is no compelling reason for humans to even serve as explorers."

Pyne points out that a flood of scientific data remains to be harvested on other planets, but this will not be sufficient motivation:

Where is the compelling *human* interest in such discovery? Where is the excitement, the passion that ultimately comes from moral drama? How can great literature and art come from such nonencounters?

From where comes the requisite public enthusiasm to sustain costly ventures? . . .

That intellectuals are curious about the new discoveries, or certain arts and sciences may benefit from them, does not mean that a society will elect to pursue them. Society must have a reason to choose such discovery, to believe in it, to thrill or despair to its revelations.

Pyne goes on to argue that we can best explore by proxy in the coming era:

Humans do not have to be physically present at the discoveries of the third age, and there are good reasons for arguing that they should not be. Robots, cameras, and high-tech instruments can get to the critical environments, record the sights, and take the necessary measurements. With television, the revelations can be broadcast instantaneously and equally for all to witness. Moreover, robotic exploration can do this for far less cost and less risk than required by manned voyages.

The Case for Space

When I reread Pyne's essay, I immediately grew curious as to how well such a system would work. No humans had traveled beyond Earth orbit since 1972, but many robot missions have explored our solar system. The *Lunar Prospector* was mapping the Moon at that time, with negligible media coverage. I had encountered no story or mention of its daily progress in the television or radio news. Finally, I turned to the mission's Internet home page, to share in the latest discoveries. I received a wealth of backup information but only the following current bulletin: "The Lunar Prospector spacecraft continues to perform very well, and all instruments continue to collect good data, according to Mission Control at NASA's Ames Research Center." So much for the instant revelations.

Subsequently, the *Lunar Prospector* did detect ice at the poles, a significant finding. But the spacecraft did not share the moment of discovery with the public. The robot *Prospector* also failed to report its reaction to the sight of sunrise when observed from lunar orbit. Fortunately, Harrison Schmitt, of *Apollo 17,* was able to describe what he had seen a generation earlier:

We looked first toward the airless sunrise and saw increasingly bright and glowing streamers of the corona radiating vast distances from a hidden sun. A faint, thin, glowing halo, possibly from dust just above the lunar horizon, joined each streamer and faded gradually to the north and south of our orbit. Then, just before sunrise, the center of each streamer became a blade of brilliant light projecting above the dark lunar horizon like a luminous sword. A few high peaks cast their reflected light from the horizon's edge across the dark highlands and then—sunrise!

Robots are not eloquent, nor are their achievements celebrated in the media. Their missions are cheaper than manned ones, but they can also fall under the budget axe. If we accept the idea that only "meaninglessness" lies out in space, then why send anything at all? Different answers have been put forward by competing philosophies that seek the purpose of human life.

For those who accept the teachings of the Bible literally, our existence centers on Earth. It is here that we perform the moral acts that determine our salvation. In the not-too-distant future, history will end with Armageddon and the Last Judgment. Space exploration, while not sinful, is an expensive diversion, one whose costs should be minimized or eliminated.

The pessimists, who I have termed the Sour Lemon School, reach similar conclusions from different assumptions. They see the universe as a barren, hostile place in which life has been produced on one planet by an unlikely turn of random events. Sooner or later, this anomaly will be erased by another random catastrophe. Prescriptions as to how we should behave in the interim may differ in detail, but the gloom surrounding our situation saps the energy from any proposed long-term effort.

A third alternative flows from the idea of Cosmic Evolution. Life and intelligence are seen as inevitable products of the laws of this universe. Individual species or entire biospheres may be extinguished, but others are generated afresh. We, of course, are the recent product of a success story that has continued its unbroken 4-billion-year run. But we remain very vulnerable to extinction as long as we are confined to this single planet. Meteorite impacts, nuclear wars, or other catastrophes of types that we have not yet recognized could snuff out advanced life on Earth. Even in the absence of sudden disasters, inevitable climate changes and the deterioration of our Sun will eventually make Earth uninhabitable. If we value the experience of being human for its own sake, then we will want to safeguard what we have gained by extending our presence into the larger universe. We will also be very curious to learn what else evolution may have produced, and ultimately see if there are other cultures whose experience can enrich our own.

At present, the key question concerning the place of life in the universe remains unanswered: Is the universe fertile or barren? Possibly some lucky result in the Stardust mission or a Mars sample return may toss the answer in our laps for a limited cost. But we are totally uninformed as to how truly alien life forms may appear, and I suspect that active human intelligence at the site may be needed to ferret it out. Robots will be helpful in pointing us in the right direction, but ultimately we will have to do the jobs ourselves. In doing it, we will get used to the sight and feel of places off the Earth— an important prerequisite if we are to live there.

With the hopeful assumption that some of us feel these goals worth pursuing, we can move on to ask how best to search.

The Quest for Life

We will want to find clear evidence for the Life Principle. If we can show that life has evolved independently more than once in the same solar system, then there is likely to be lots of it in the universe at large. If life has developed under conditions very different from those on Earth, then we can assume that it started separately. If the conditions resemble those on Earth, but some feature is radically different—for example, a totally different genetic material or building blocks of the opposite "handedness"—then it is also fairly safe to assume a separate start. On the other hand, if the life that we discover has a biochemistry similar to our own, then the presumption will remain open that it has spread from one world to the other by local panspermia. This discovery would be exciting but would leave the fundamental question unanswered. This possibility must be kept in mind when we consider the various kinds of evidence that we might find for the existence of life.

Evidence for life beyond our planet could present itself in four different ways, which I will call the four E's:

1. *Existing life.* Living creatures of any kind would provide the jackpot for discovery. We could study them in depth, compare their function to our own, and learn some broader rules that govern biology in the universe.

2. *Evolving life.* We might encounter some systems that are less com-

plex than our own bacteria but much too organized to represent random chemical fluctuations. We would have caught an authentic example of the Life Principle at work. We might even learn something about our own beginnings, as all of the intermediate steps seem to have vanished on this planet. If the chemical system was an unfamiliar one, however, we might need to study it for a long time to assure ourselves that complexification was taking place.

3. *Extinct life.* We could find microfossils, as we may have done in ALH84001. If the remains were limited to simple shapes, however, then we could wind up with the same type of controversy that has followed that Antarctic meteorite. We would not learn details of the biochemistry in such a case, to understand whether the life has arisen separately. In very cold places, however, we might recover the component chemicals as well as the life forms, which would be much more useful.

4. *Extrasolar life.* Our solar system has been in place for over 4 billion years, a good fraction of the lifetime of the entire universe. Some meteorites and lunar rocks have survived from that time. If intelligent beings, or their robot craft, had visited our system for survey purposes at any time during that long interval, then they may have left behind artifacts that still survive. If so, we could search for them.

For such a search to succeed, a number of separate assumptions would have to work out. Chances of success here are much smaller than in the first three cases I have named. On the other hand, the importance of such a discovery would be so profound, and the search itself would so excite the public imagination, that it should be considered in any long-term program. The strategies for this search would differ from those we will use for the others, so I have put the discussion into a later chapter.

With our objectives in mind, we have a host of worlds from which to select. How do we choose? In seeking life, our priorities may be very different from those of a planetologist interested in the history of our solar system or an astronomer seeking to understand the behavior of the rings of Saturn.

We can start by noting that life is thriving on our own planet. We are well endowed in terms of the requirements for chemical evolution set out in the chapter on the Life Principle (Chapter 6). On Earth we have had surface oceans, an atmosphere that exchanges material with them, and plentiful energy supplies (solar energy, volcanoes, wind, moving water, thunderstorms, and more) for billions of years. What about the other worlds? They have had as much time as we have had here. We do not know which sets of chemicals will work and which will not, nor do we even know which ones started life here, so we cannot say much on that topic. The easiest requirement to monitor is the need for a liquid or dense gas, which is needed so that reactions can take place readily.

We cannot be sure that any body in our solar system other than Earth has a long-lasting liquid on its surface. But we can look for atmospheres, which may or may not come with oceans. Seven planets, including our own, have appreciable atmospheres. That list includes the six largest planets in size and the ninth, Titan, a moon of Saturn. Obviously, bulk helps in retaining gases.

Five of the seven reside in the outer solar system, where planets coalesced far from the heat of the Sun. They were able to retain lighter gases such as hydrogen, helium, methane (the simplest compound made of carbon and hydrogen), and ammonia. Four of them, Jupiter, Saturn, Uranus, and Neptune, succeeded so well that most of their bulk is made of such materials, surrounding a small rocky core. They are often called "gas giants," though most of their light materials are under high pressure in the interior and exist in another, nongaseous, form. The fifth, Titan, is interesting enough to deserve a separate treatment in a later chapter.

The other two worlds with atmospheres, Venus and Earth, orbit the Sun at a closer distance. (Mars has a skimpy atmosphere, less than 1 percent that of Earth, but it has other reasons to be of great interest. We will award it a separate chapter of its own.) The lighter gases mentioned previously have been lost, but heavier ones such as nitrogen, carbon dioxide, and (on Earth) oxygen are present.

Venus is worth a detour. It had been regarded as a sister world to Earth and a likely place for life of our own kind, because its clouds shielded it from telescopic observation. Orbiters and landers have examined the terrain, and perhaps nowhere else have our dreams and the reality differed so sharply. We can start with the adventures of a fictional traveler, Ham Hammond.

A Fine Winter Day on Venus

Mid winter . . . in the Venusian sense . . . is nothing at all like the conception of the season generally entertained on Earth, except possibly by dwellers in the hotter regions of the Amazon basin, or the Congo.

They, perhaps, might form a vague picture of winter on Venus by visualizing their hottest summer days, [and] multiplying the heat, discomfort and unpleasant denizens of the jungle by ten or twelve. . . .

In the winter, the temperature drops sometimes to a humid but bearable ninety, but, two weeks later, a hundred and forty is a cool day near the torrid edge of the zone. And always, winter and summer, the intermittent rains drip sullenly down to be absorbed by the spongy soil and given back again as sticky, unpleasant, unhealthy steam.

And that, the vast amount of moisture on Venus, was the greatest surprise of the first human visitors; . . .

The situation made eating and drinking in the open a problem on Venus; one had to wait until the rain had precipitated the spores, when it was safe for half an hour or so. Even then the water must have been recently boiled and the food just removed from its can; otherwise . . . the food was apt to turn abruptly into a fuzzy mass of molds that grew about as fast as the minute hand moved on a clock. A disgusting sight! A disgusting planet!

The molds described were far from the most voracious creatures depicted in this 1934 story, "Parasite Planet." My favorite was the doughpot —a mass of "mushy filth" that crawled amoeba-like, devouring everything in its path. The hero of this epic, Ham Hammond, had to overcome a host of monsters, including doughpots, trees that tossed lariats to catch their prey, others that swung spikes instead, and hostile three-eyed, white-claved Venusian natives en route to meeting the lovely human heroine with whom Hammond fell in love.

The author was Stanley G. Weinbaum, who, according to Isaac Asimov, "had been instantly recognized as the world's best living science fiction writer." Unfortunately, Weinbaum died of cancer the next year, at age thirty-three. His masterpiece, "A Martian Odyssey," was set on that world, but no planet of our system was safe from his fertile imagination. His imaginative and menacing creatures dominated "Parasite Planet," but his basic picture of Venus was one that many science fiction writers and some scientists shared until the mid-twentieth century.

Venus is only slightly smaller than the Earth, and its 225-day year is not too different from our own. As it orbits the Sun at just under three-quarters of our own distance, we would expect it to have a warmer climate. Further, it lies wrapped in thick clouds that prevent direct telescopic observation of its surface. These arrangements both suggested a humid place and prevented any easy contradiction. Thus Nobel Laureate chemist Svante Arrhenius (we have met him in connection with panspermia) could claim the following in 1918:

> We must therefore conclude that everything on Venus is dripping wet. . . . A very great part of the surface of Venus is no doubt covered with swamps. . . . The temperature on Venus is not so high as to prevent a luxuriant vegetation. . . . Only low forms of life are . . . represented, mostly no doubt belonging to the vegetable kingdom; and the organisms are nearly the same all over the planet. The vegetative processes are greatly accelerated by the high temperatures.

Even more optimistic views were put forth by Charles Greeley Abbot, the director of the Smithsonian Astrophysical Observatory, who also thought that Venus might hold intelligent life: "A twin planet to the earth in size and mass, its high reflecting power seems to show that Venus is largely clouds indicative of abundant moisture; probably at almost identical temperatures to ours, our sister planet appears lacking in no essential to habitability."

When scientists had the opportunity to peer beneath the clouds in the late twentieth century, however, they learned that nature had other plans in mind. Orbiters of the Soviet Venera and American Pioneer and Magellan series circled the planet and mapped it by radar. Venera probes parachuted through the atmosphere and transmitted photographs of the surface. The pictures that we obtained differed radically from the early dreams.

Venus Today: Hotter Than Hell and Nothing to Drink

Weinbaum's adventurer, Ham Hammond, would face problems more harrowing than the doughpots if he found himself on the real Venus. At a searing 460°C (860°F), the surface temperature has a quality that makes comparisons to earthly jungles ludicrous. The atmosphere, carbon dioxide with a lacing of nitrogen, would not only be unsuitable for him to breathe, but its pressure, ninety times that of Earth (the equivalent of more than a kilometer of sea water), would crush him. To put it mildly, he would need a very well constructed space suit.

If Hammond stood where the last Venera probe had landed, he would see a gray, flat, lava-carved plateau with the texture of cracked ceramic tiles. The light would be dim and diffuse, with a deep red cast that was caused by atmospheric scattering. The clouds themselves would be distant, concentrated in a layer some 50 kilometers above the surface. If he remained where he was, he would eventually experience nightfall, as day and night on Venus each last about two Earth months. Then the sky would grow dark, but his environment would be lit by the red-hot glow of the ground and occasional lightning flashes.

Hammond might choose to travel: Venus lacks oceans, and its land area is therefore much larger than that of the Earth. If he did, he would cross an uneven terrain carved mostly by the action of volcanoes, with some features due to meteorite impacts, and traverse deformed volcanic plains studded with steep-sided pancake domes. In places he would cross flat lava-flow "oceans" fed from channels, which were in turn fed by lava "rivers." If his journey was long enough, he would encounter many novel and irregular features: spiderweb fracture patterns, circular fracture rings, huge cliffs, and vast areas of cracked and wrinkled terrain. Possibly, active volcanoes still function on Venus, and if so, he might get to see a lava eruption.

Unfortunately, he could not expect a break in the heat wave or (with one exception) any improvement in the climate in his travels. The atmosphere redistributes the heat and keeps the temperature of Venus largely the same, from equator to pole. The only direction that he could go, if he wanted to escape the heat, would be up. The mountains on Venus rise to about 11,000 meters, higher than Everest. Near those heights the temperature falls to "only" 440°C (820°F). To really cool off, Ham would have to fly upward into the clouds. At midcloud levels the pressure would be only 70 percent of that on Earth, with a temperature of 42°C (107°F). Finally, at the cloudtops, a blissful Earth-like 13°C (55°F) would be attained. But Ham should not open the windows of his plane and take off his helmet. Those clouds are not made of water but rather of concentrated droplets of corrosive sulfuric acid, with perhaps some sulfur giving them a yellow color.

Complexification on Venus?

We know life only as we have seen it on this planet. Our first inclination when considering other worlds has been to extend our experience. We saw

Venus as hot but habitable, with creatures that were unpleasant but shared enough Earth biochemistry to digest us without complaint. When these visions collapsed, astronomers rushed to firm conclusions; for example, "extant life on Venus is out of the question" and "it is unlikely that life ever arose on Venus."

If we wish to search for a Life Principle, however, we cannot be so abrupt. Water is not unique—other liquids can support chemistry. Venus, alas, has no oceans of any type on its surface, except for possible transient lava flows. A dense atmosphere with liquid droplets, however, may support chemical cycles leading to self-organization. Energy poses no problem: ultraviolet light, chemical sources, and other energy supplies are available in abundant amounts. One planetoligist, David Grinspoon of the University of Colorado, has been willing to speculate along these lines.

Grinspoon cites novel features on Venus that indicate that something that we do not understand is going on. The clouds contain dark mobile markings, visible in the ultraviolet, and unusual particles. The mountain peaks are all "surprisingly shiny." That glitter is not due to ice but to some mineral substance, such as pyrite or tellurium. Possibly, complex cycles based on sulfur chemistry are at work and some form of complexification is taking place.

My late collaborator, Gerald Feinberg, had speculated earlier about life in the clouds of Venus. For organisms adapted to the cloud temperatures, descent to the furnacelike surface might be a disaster. Survival for them would depend on buoyancy, and they would need to take advantage of updrafts or use devices such as airbags filled with lighter gases. Evolved creatures might have such resources, but it is hard to visualize an extended series of earlier steps taking place in suspension above the inferno. One other scenario remains possible, however.

The current surface of Venus appears to be only a few hundred million years old. Some catastrophe involving massive lava flows may have eradicated a different set of conditions that existed before that time. If early Venus was more Earth-like in nature, then any creatures that evolved in that era may have had time to adapt and flee to the clouds, before disaster caught up with them.

The extended series of assumptions and speculations needed to imagine cloud life on Venus does not make it our best bet, and the corrosive nature of the chemicals would hinder robotic exploration. Sooner or later we should look there, but we will find better possibilities for cloud life in the gas giants. Each of them has its own distinctive features, but I will select only one, the closest and largest, for discussion here. If it should meet our expectations, then surely we will place the others on our agenda.

The Bulk of the Solar System

Our system does not follow a policy of "share and share alike" with the material that it holds within it. Jupiter is definitely the majority stockholder; it hoards over two-thirds of the matter outside of the Sun. To get an idea of Jupiter's size and appearance, we will return to our Museum of the Cosmos.

As you may remember, a model Earth at Level 7 was just a bit less in diameter than our height and the Moon was a beachball, about half a city block away. We could not put a Jupiter model into the same room, if we kept to that scale, as it would be 63 kilometers away at closest approach. Instead, we will give it a display area of its own. We will want one with a high ceiling, as the model would be 15 meters tall. We will decorate it gaudily with orange, red, white, and brown stripes so that it resembles a large helium-filled excursion balloon. In most balloons that I have seen, the stripes run vertically, but our Jupiter stripes will orient horizontally, aligned with its rotation and orbit. To extend the decoration we will add a prominent reddish spot, larger than ourselves, to the bottom half of the model. Fittingly, astronomers have named it the Giant Red Spot. We will want to animate this model to complete the picture. Alternating stripes move with the planet's rotation, with features such as the spot coming to the same place every ten hours. Our completed model will be much more entertaining than one of Venus, which would be nearly uniformly white.

In the real Jupiter, the Giant Red Spot represents a cyclone, a violent circular windstorm that has been observed for 330 years. The colored zones are caused by persistent flowing bands of wind that differ in their chemical composition. We do not know the identity of the substances that cause the colors. Occasionally, smaller disturbances, or eddies, may arise and interact with the overall pattern before dissipating. Planetary scientist Andrew Ingersoll has described the situation: "If one regards the jets as an ordered flow and the eddies as chaotic, observations suggest that (in this case) order arises from chaos."

This statement should immediately put us on the alert. Self-organization has taken place to some extent in the surface layers over the entire planet.

What may be going on inside? Our current information is sketchy at best, but we can put it into perspective by picturing a parachute descent into the atmosphere of Jupiter, using an imaginary ship of our own design. We will be following the precedent of a real probe that was dispatched by the *Galileo* orbiter, entered Jupiter's atmosphere on December 7, 1995, and reported data for fifty-eight minutes before it was destroyed, about 160 kilometers deep within the atmosphere. However, the actual probe descended at an atypical place, a hot spot region dubbed the "Sahara of Jupiter" because of its relative dryness and lack of clouds. We will select a more normal location for our excursion.

We would first see blue sky above the cloudtops, as we do on Earth, before we descended into wispy white clouds of ammonia flakes. We would find it quite cold outside, with a temperature near $-123°C$ ($-189°F$), and the pressure much less than that on Earth. We could not confuse our location with the clouds of Venus, as a very different set of gases dominates the Jovian environment, one much more in line with the overall composition of the universe.

Hydrogen gas, made of the simplest element, represents about three-fourths of the Jovian mix, with helium, the second simplest one, taking up much of the remainder. Jupiter contains some ammonia, a feature that makes its atmosphere very different from Venus's, which had clouds of sulfuric acid. The presence of methane, the simplest compound of carbon and hydrogen, also sets Jupiter apart from Venus. The latter planet has carbon dioxide (carbon bound to oxygen) as the major constituent in its atmosphere.

As we descended farther in, we might encounter impurities in the ammonia clouds that afford the colors, though they might not look as striking at close range. Red and white would gradually give way to brown and then blue as we fell, while the temperature grew warmer and the pressure higher. Jupiter produces heat internally by gravitational compression, so the heat would go up and up as we went down and down. If the planet had been seventy times more massive, it would have ignited by thermonuclear fusion and become another Sun.

In continuing our descent, we would pass through another layer of clouds made of ammonium hydrosulfide (a combination of ammonia and the noxious gas hydrogen sulfide) and then reach a possible region of clouds of ice and water droplets. At last we would reach a comfortable place, with temperatures approaching those on Earth's surface. But we would be in the dark because of the clouds above, and ferocious winds would keep our nerves on edge. We still would have covered only about a fifth of a percent of the distance to the center, which lies 71,000 kilometers below the cloudtops.

If we have been wise, we will have equipped our imaginary ship with an engine as well as a parachute, for we would not want to travel farther down. By the time we had penetrated 20,000 kilometers in, the pressure would be 4 million times that on Earth, and the temperature near 10,000°C. The hydrogen around us would gradually have converted from a gas to a liquified state, and then to an unusual liquid metallic form. These conditions fall outside the range that we can study in laboratories, so it is difficult to predict what our environment would be like. We would have to descend through this strange medium before we hit an Earth-sized chunk of rock at the core.

Rather than take this trip, we will do better to cruise through a region with earthly temperatures. Only lightning flashes would light the way (thunderstorms are always in season on this planet), so we will want to provide our own illumination. In this adventure, we would have much in common with the submersibles that search for life in the darkness at the deep sea bottom. What surprises might await us in the dark clouds of Jupiter?

Several writers have wondered whether life could evolve and exist there. The environment contains the very gases that were used in the Miller-Urey simulation of the early Earth: methane, hydrogen, ammonia, and water. If any organisms did arise, they would suffer from continual danger, due to the absence of any firm support. No barrier would exist to protect evolving chemicals or cells from being swept into the depths and incinerated below. Two decades ago, Carl Sagan and E. E. Salpeter estimated that bacteria-sized creatures might survive for one to two months before they drifted far enough down to be destroyed. If they reproduced rapidly and took advantage of updrafts, however, these "sinkers," as they were called, might survive.

Much larger "floaters" were also imagined that kept at a constant level by devising balloons, which could be filled with pure hydrogen or perhaps hot Jovian air. These creatures would either scavenge the cloud environment to collect their food and the building parts for their bodies or else manufacture their own supplies at the cloudtops, using sunlight for energy.

Yet other creatures, the "hunters," would prey on the floaters. This jungle-in-the-clouds is portrayed in the illustrations for Carl Sagan's book

Cosmos. In one scene, a group of floaters clusters near an updraft while camouflaged hunters prepare for an attack.

How shall we react to such colorful and detailed speculations? NASA scientist R. D. MacElroy took a dim view of them in a review article from the same era:

> Postulating surfaces with stability in a turbulent atmosphere requires imagination and faith, but there is historical precedence for the use and overuse of these qualities. . . . Inevitably, it is harshly revealed that historical descriptions of such magnificent beasts as unicorns and mermaids have been the result of naïveté and the absence of facts. A biologist considering the possibilities of life on other planets and their satellites is the most naïve of scientists. And he is often uncomfortable when making such considerations, because the basic facts he needs are not available.

No physical laws have been violated, however, by Sagan's imaginative ideas. Science does not advance through negative judgments based on philosophy. When basic facts are lacking, then the most straightforward remedy is to collect them.

We can hope for a future expedition in which our imaginary craft has become an authentic automated Jupiter cruiser, with floodlights, a microscope, and a chemical analyzer aboard. And as Stephen Pyne suggested, we should include television cameras to return the images, as our vessel cruised the clouds. Even if no hunters should stalk us, or floaters drift within our beam, there would still be much for us to learn.

Turbulent downdrafts and scarcity of key ingredients may well have prevented the evolution of recognizable life forms on Jupiter, but complexification may have weaved its web in another way.

As we have seen, atmospheric processes have organized themselves on a global scale, and chemical processes may have done so as well. The entire planet would then be in a state of partial organization, a unified entity that has not subdivided into individuals. We could learn of this by studying the chemical complexities and differences from one layer to another, and perhaps by comparing Jupiter to the other gas giants. We might not choose to call this life, but it would still allow us to see the Life Principle at work.

Life below the Crust

We have considered planets whose atmospheres are thicker than our own. At the other extreme, we find a large number of airless worlds. We should not pass all of them by: some of them may still serve as havens for life if they hold liquids beneath their surface. Once again we will find the larger worlds more promising than the tiny ones. The big ones have a greater internal area compared to their surface and can hold onto their original heat of formation for a longer time, as the smaller ones go cold and dead.

Another circumstance, in addition to sheer bulk, will help keep some worlds warm inside. Those satellites that orbit a giant planet at close range are subject to tidal forces from the larger world and experience its magnetic field as well. The tides, in particular, may heat up a world that would otherwise have frozen solid.

One internal fluid that we know well from our own planet is molten rock. We cannot rule this out as an arena for self-organization, but we have only observed it on the surface, in the process of turning solid. To examine it in its native state we would have to dig into the deep interior of our own world. This effort might bring up more technical problems than a close-up study of the surface and near-surface regions of other worlds. We will deal with more familiar liquids first.

Similar arguments apply in the case of Io, the innermost moon of Jupiter. That world has informed us, by sending up volcanic eruptions, that it has fluids inside. The spectacular release of extremely hot molten lava and sulfur dioxide gas was recorded by passing spacecraft. Lakes of sulfur and molten rock and perhaps ones of liquid sulfur may exist beneath a 1,000-meter crust on Io. Self-organization may be possible in this medium, but our unfamiliarity with its chemistry may again make it difficult for us to recognize it.

A more tempting target is Io's next-door neighbor, Europa. The crust in this case is ice. No water eruptions have been seen so far, but the plastic, relatively uncratered nature of the crust suggests strongly that a water ocean lies beneath. We will flag this one as very promising and devote part of a chapter to it.

Enceladus, a small moon of Saturn, represents a similar case. A large portion of its surface is smooth, shiny, and uncratered; this suggests that the surface melted rather recently. An ocean may lurk inside, though it is not clear what energy source has melted its interior. Europa, because of its proximity and larger size, will undoubtedly claim priority for study.

Other worlds of interest could be given an honorable mention. For example, Triton, a large moon of distant Neptune, has a surface that is extremely cold even by solar system standards, yet it gives out plumes of smoke and has a very thin atmosphere of nitrogen and methane. But we must not extend ourselves too much. When we search, we may find what we want quickly. More likely, the Life Principle will only yield its secrets to patient and prolonged investigation. Planets may be compared to human friendships in that an intimate relationship with a few can be better than surface contacts with a horde.

I have chosen three worlds for discussion in the next chapters. Mars has a possible past history of life and a chance that some remnant still survives now. Europa offers an internal water ocean, and Titan provides its thick atmosphere with the possibility of organic liquids on its surface. These three appear to have been given priority by NASA as well. They also seem appropriate for future human exploration, as they offer sky, ocean, or both, as well as ground underfoot. We need some gentle transitions in our voyage out there; the more alien environments can wait for their turn.

9

The Big Orange

As we saw in the first chapter, early dreams of extraterrestrial life centered on the Moon. But as the nineteenth century wound down, telescopic observations made it clear that the Moon was an airless, desolate body. Mars replaced it as the focus for human dreams about extraterrestrials. Unlike the other planets, which, like Venus, were wrapped in clouds or too distant for detailed observation, Mars offered a surface that could be examined by telescope. In 1877, the Italian astronomer Giovanni Schiaparelli observed the planet on a favorable occasion and reported linear markings, which he named *canali* (the Italian word for channels). Two decades later, American astronomer Percival Lowell interpreted them as artificial canals, the work of intelligent beings. The grand era of Martian fantasy had begun, and continues to this day.

The prominence of Mars both as a favorite subject for scientific observation and as the provider of all manner of creatures for science fiction novels, stories, and films has put it at the center of the debate on extraterrestrial life. The advocates of various philosophies concerning the place of life in the universe have seized on every bit of suitable evidence and enlisted it for their causes. In the continuing cross fire of charges and countercharges, it has become difficult to learn what the planet actually can tell us about the past and present existence of life. One culminating event in the ongoing debate was the successful landing of two Viking landers equipped with life-detection experiments on the surface of Mars. This engineering triumph was carried out in 1976, almost 100 years after Schiaparelli's observations. The results of life detection, however, were ambiguous and confusing. We can sample, however, how they were interpreted by the schools of philosophy that have followed us on our travels.

We will let Duane Gish speak for the Creation scientists, adherents to the cosmos as described in the Bible:

> Life could not have arisen spontaneously on this earth by any combination of natural processes that we have operating today. Our experience on Mars supports that conclusion. On Mars we have a flow through of energy. The evolutionists say that is all that is needed, just a flow through of energy and order and complexity will be created. We have those conditions on Mars, but we have found absolutely nothing, no life, no organic molecules of any kind, precisely as predicted by creationists. Carl Sagan, on the other hand, had predicted we not only would find life there, but that we would find large forms of life, something we could see even without a microscope. We found no life, however, and no organic material of any kind on Mars. Why not? If you want a natural test of theories on the origin of life, that was it, without any scientists, creationist, evolutionist, or otherwise, imposing conditions on the experiment. We did not find life; we found nothing.

For the pessimistic group that I have named the Sour Lemon School, we will hear Dr. Norman Horowitz, retired professor of biology at the California Institute of Technology. Professor Horowitz was responsible for one of the three life-detection experiments on the Viking landers.

> Viking found no life on Mars, and, just as important, it found why there can be no life. Mars lacks that extraordinary feature that dominates the environment of our own planet, oceans of liquid water. It is also suffused with short-wavelength ultraviolet radiation. Each of these circumstances alone would probably suffice to ensure its sterility.
>
> The failure to find life on Mars was a disappointment, but it was also a revelation. Since Mars offered by far the most promising habitat for extraterrestrial life in the solar system, it is now virtually certain that the earth is the only life-bearing planet in our region of the galaxy. We have awakened from a dream. We are alone, we and the other species, actually our relatives, with whom we share the earth.

A somewhat different interpretation of the results was produced by

Dr. Gilbert Levin, the principal investigator for another of the Viking life-detection tests, the labeled release (LR) experiment:

> Each of the reasons supporting a non-biological interpretation of the LR Mars data has now been shown deficient. The demonstrated success of the LR in detecting microorganisms during its extensive test program with its record of no false positives can no longer be denied. New evidence, together with the review of the old, leaves the biological explanation standing alone. The scientific process forces me to my new conclusion: the Viking LR experiment detected living microorganisms in the soil of Mars.

Let me add my own point of view: We do not know whether life exists on Mars right now, or did so in the past. If any life should exist, it probably is microscopic, as nothing visible has been spotted by cameras in three locations. Despite the Gish and Horowitz statements, Mars has only a thin atmosphere and no liquid on the surface, so it is not the most favorable place in the system to test the Life Principle. As the most Earth-like planet in our system, however, it has captured our imagination, and it may provide a home for humans in the future. In recent NASA photographs, it looms orange-pink, with mottled brown markings, which prompted me to call it "The Big Orange." It has certainly earned our attention. To follow its evolution in our thoughts (and with apologies to Dickens's *Christmas Carol*), I have devised visions of Mars Past, Mars Present, and Mars Future. (The first and last are set in the guise of science fiction.)

Mars Past: A Martian Fantasy

Captain John Cooper looked through the small porthole of his ship, hoping to get his first close view of the surface of Mars. He had been too occupied in maneuvering his vessel on the previous evening to take in much of the surroundings. All had gone well until the sudden dust storm swirled up, blocking his vision. He had lost sight of the canal intersection that was his landing target, and he hunted for almost an hour until a break in the clouds afforded him a clear view of open ground. He needed all of his skill as a bomber pilot in World War II to bring his craft down safely. By that time the Martian sunset had winked out, and he fell asleep exhausted.

The ship had performed well during the boring months-long journey from Earth. It was the first of its type to be constructed and had not been tested, so Cooper had not known what to expect. He had built it himself on his New Mexico ranch, with the help of volunteers from the local high school and using parts smuggled to him by friends who were still in the military. The students and curious passersby had been told that it was a model, to promote tourism and to stimulate interest in space travel. That was exactly what it would have been if he hadn't had the plans for the rocket drive. Major Cooper had been a member of the intelligence team that had raided the Nazi V-2 base at Peenemünde in the closing days of World War II. They had looked for evidence of a German atomic bomb. That had been a false lead, but while there, John had had the chance to copy the plans for the engine that had brought him to Mars. Later, after the war, he conferred with Wernher von Braun concerning the details.

Major Cooper knew that the U.S. government would eventually get around to the exploration of Mars. He also knew that they would take many years before they decided to do it. Probably, they would want to go to the Moon first, and John knew that that was a boring chunk of rock. Mars was different, and he wanted to set an example by getting there on his own. As a boy, he had been thrilled by Percival Lowell's reports of a dying Martian civilization, one that struggled to survive in a desertlike planet. The civilization had succeeded by building a network of canals to bring water from the Martian polar regions to temperate areas for irrigation. In fact, Major Cooper had used Lowell's maps to navigate in close to his target, near Ascraeus Lucas, a prominent canal intersection. But after the storm, he had no idea where he had landed. As he looked out, he saw a desert with rolling hills in the distance, much like New Mexico. But no canal was in sight.

Pinpointing his location was only one of his problems. Due to weight restrictions on takeoff, his rocket ship had only carried enough food and fuel for a one-way trip. He would have to forage for supplies and synthesize his fuel if he ever hoped to see his home again. Of course, those tasks might not be possible unless he could find some cooperating Martians. H. G. Wells and others had depicted the Martians as hostile, but Cooper felt that any truly advanced civilization was bound to be helpful. He had especially admired the stories of Edgar Rice Burroughs and hoped that he might meet someone as beautiful as Deja Thoris, the Princess of Helium. But he was also a practical man. He had brought rifles, hand grenades, and even a small mortar, just in case things turned out badly. But he could only check out his situation if he explored; he would have to go out. Opening the airlock of his ship, he climbed out to face the Martian morning.

The air was cool and thin, but no worse than he had encountered when he had climbed the Matterhorn. He took in deep lungfuls, finding it quite breathable, and noted only a slight spicy aroma. The ground was covered with a yellow mosslike vegetation that he couldn't identify, and occasional scrubby short bushes. His stomach gave him warning pangs, and after months of Spam and army rations, fresh meat was a high priority. He decided to deploy his Rover to search for it. Rover was his hound dog, his trustworthy companion on many hunting expeditions in the hills near his ranch. John had felt that Rover would prove more useful than any human companion whom he could bring along, and weigh a lot less. He also had visions of the article that he would sell to the *National Geographic* or the *Saturday Evening Post* on his return: "Man and Dog on Mars."

He gave the dog a signal to find food. John watched as Rover stumbled around for a few minutes, befuddled by the low Martian gravity. He quickly adjusted and bounded off into the distance, kicking up a huge cloud of dust. A few minutes later he returned, dragging a dead squirrel-sized animal of a very unusual appearance. Its face was rodentlike, but it was hairless, with three legs on either side and a broad flat tail that was wider at the tip than at the base. However strange the beast may have appeared, it tasted quite nice, much like chicken, after it had been roasted over a fire made from collected dead twigs and branches.

With his hunger now appeased, Cooper put a canal hunt at the top of his priority list. His compass was of no help so he simply set off, with Rover by his side, in the direction where the land sloped gently downward. After some practice, he got quite accustomed to the low gravity, in which he could move along in leaps and bounds of up to 50 feet, if he chose to do so. He settled for a quick trot and was rewarded within an hour when he saw a canal, running straight as a ruler into the distance. Its pale green waters shone as smooth as polished glass beneath the pinkish Martian sky. A band of low vegetation, perhaps an unfamiliar crop of some sort, ran alongside the canal.

Most interesting of all to Cooper was the asphalt road that separated the vegetation from the desert. The dusty soil continued to its edge and then stopped, as though the road were coated with a repellant of some sort. As Cooper weighed which direction to choose in continuing his journey, he suddenly became aware of a speck in the distance, approaching along the road at high speed. He motioned Rover to be quiet but alert and reached for his rifle.

Cooper was made of the right sort of stuff, as space travelers go, but the next events made him doubt his sanity. Only the calm and somewhat bored behavior of his dog reassured him. The vehicle approached to within 100 yards and halted. Two figures emerged and slowly moved toward him. Cooper would not have been ruffled by anything exotic, but the vehicle was an open Ford convertible of recent vintage, and the leading person was dressed in a dark pin-striped banker's suit. He was waving a white flag. The second individual was also in pinstripes, but those of a New York Yankees baseball uniform. He was carrying an American flag rather than a bat.

It was only when they drew close that Cooper became sure that he was not hallucinating visions of home. Both men were short, about the size of average twelve-year-old boys, and their complexions had a definite olive-green cast. Their hats were pulled on tightly, but he thought that their ears were definitely pointy. Finally, a mysterious bulge thickened their midsections. It could have concealed an extra pair of limbs.

His unease calmed down somewhat when they spoke to him in fluent, if gutteral, English. The banker introduced himself as Morris Thoris, the Baron of Boron; the baseball player was Tars Kajak, the Martian home run champion, but he preferred to be known as the Tharsus Thumper. Their language, car, clothing, and sports had all been copied from Earth and particularly American television programs. Before that they had paid rapt attention to our radio broadcasts. Martian interest in these programs, particularly comedy and sports, was phenomenal. Though none was in sight at the moment, enormous television signal receivers now dotted the landscape. The Martians had even placed an unobtrusive satellite in Earth orbit to amplify the signals and beam them to Mars. Every Martian schoolboy had learned the songs of the Andrews Sisters and followed the heroics of the Yankee Clipper. The Milton Berle comedy show was watched by a phenomenal 70 percent share of Martian households.

These developments had come just in time to revive Martians' stagnant civilization. Martians had lost interest in life. Canals, roads, and factories were falling into disrepair. The young people had turned away from science and technology. Now they were thoroughly Americanized. Through us, they had rediscovered their sense of humor and of play. They were especially excited to have a visitor from Earth. Cooper's voyage had been monitored through his radio broadcasts, and they were prepared to give him a hero's welcome, including, of course, a tickertape parade. If he and Rover would care to step into the car, they could proceed to the city.

While they were en route, the Baron confessed that he had prepared an additional surprise. An American service station appeared by the road with a very familiar logo. As they pulled up, a group of Martian attendants dressed in American uniforms linked hands and began the song that he expected. Cooper had watched that particular TV show and gritted his teeth, for he knew what was coming. It's time for your *make-up* shouted the Baron as an overstuffed sack of flour was puffed into Cooper's face.

Author's Explanation: A Martian Fantasy

I could not resist having some fun with the Martians, and many others have preceded me. I have borrowed some details from Edgar Rice Burroughs's *A Princess of Mars*, the first of a multivolume series describing the adventures of Captain John Carter on Mars. The idea of family-sized Martian rockets was borrowed from *The Martian Chronicles* by Ray Bradbury. In that volume, Martian civilization is extinguished by contacts with our civilization. I have simply reversed the plot. American culture is received with enthusiasm across the globe today. I imagined that it could also leap across interplanetary space in a type of cultural panspermia.

The comic aspects of human-Martian encounters have been more represented in films than in literature (for example, the satiric *Mars Attacks*, in which small green Martians with large brains are cast in the role of aggressors). In the early 1950s, when my own tale is set, *Abbott and Costello Go to Mars* reflected Hollywood's attitude toward the planet.

Martians were taken quite seriously earlier in the century, due to Lowell's advocacy of his canal-building civilization. The following quote from a *Wall Street Journal* article of 1907 (provided by an Internet source) illustrates their impact.

> The most extraordinary development has been the proof afforded by the astronomical observations (showing) that conscious, intelligent human life exists upon the planet Mars. . . . Dr. Lowell, director of the Lowell Observatory in Arizona . . . gives a number of photographs taken of Mars. . . . He sums up the testimony of these photographs by

saying that they reveal to laymen and astronomers that markings exist on Mars which are, of course, the lines of the great canals on Mars constructed for the purpose of irrigating that globe.

Improved telescopic observations and eventually Mariner fly-bys revealed that Lowell's canals had been an optical illusion. Despite the detailed maps that he produced, what he observed was an effect that occurs when the human eye, struggling at the limit of observation, connects irregular lines into a pattern.

After hope faded for intelligent Martians, some scientists supported the idea of extensive vegetation on Mars, citing seasonal color changes as evidence. These effects are more likely due to dust storms. Such ideas were finally put to rest by the Mariner fly-bys, which showed no sign of vegetation. In particular, the July 1965 *Mariner 4* mission, which photographed 1 to 2 percent of the planet at close range, revealed a Moon-like landscape, rich in craters. The pessimists extrapolated this vision to the entire planet and declared it lifeless. A front page article in the *New York Times* started with the words, "A heavy, perhaps fatal blow was delivered today to the possibility that there is, or once was, life on Mars." The article emphasized the absence of evidence for river valleys or other signs of water erosion. It further declared that Mars lacked volcanoes or other mountains. Mars was as dead as it was red.

The observations of the next decade showed this description to be as fanciful as Lowell's canal world. Mars has abundant evidence for features created by water and volcanoes that dwarf those on Earth. A later NASA report clarified the situation:

> Soon it became apparent that almost all generalizations about Mars derived from Mariners 4, 6, and 7 would have to be modified or abandoned. Participants in earlier flyby missions had been victims of an unfortunate happenstance of timing. Each earlier spacecraft had chanced to fly by the most lunar-like parts of the surface, returning pictures of what we now believe to be primitive cratered areas. It was almost as if spacecraft from some other civilization had flown by Earth and chanced to return pictures only of its oceans.

Telescopic observations and fly-bys could take us only so far in our understanding. If we really wanted to learn about the possibility of life on Mars, we would need a closer inspection.

The Vikings Have Landed

In 1976, robotic Viking landers settled down at two widely separated locations on the surface of Mars. Each lander had traveled together with an orbiter. Safety considerations eventually dictated the choice of sites. The selection of a suitable landing place for the first lander was delayed for weeks after arrival while the orbiter photographically mapped the candidate sites. The elaborate Viking landing strategy involved a heat shield, a huge main parachute, retro rockets, and descent engines, which slowed the descent to a crawl and let the vehicle settle gently at 7.2 kilometers per hour on its three outstretched legs. There were still elements of risk, as an insignificant rock, half a meter tall, could damage the lander's belly, and a larger boulder under one of its legs could cause it to topple over. A computer, without human intervention, was in charge of the last stages.

On July 20, 1976, *Viking 1* settled in an area called Chryse Planitia, on a vast ancient plain. When the lander touched down safely, the scientists in charge wept with joy. On September 3, the second Viking lander arrived without problems on another flat plain almost halfway around the planet. The search for life could now begin.

The two landers were duplicates, and each carried an entire science laboratory, with three separate experiments dedicated to life detection. Both landers returned similar data. As we shall see, the life-detection results were surprising and provocative. The cameras, however, saw only barren deserts, and the gas chromatograph-mass spectrometer, an instrument designed to identify organic compounds, found none. Pessimism carried the day once again, as reflected in this comment in the *New York Times* by the gifted science writer Dr. Lewis Thomas: "Mars from the look we've had at it thus far is a horrifying place. It is, by all appearance, stone dead. It is surely the deadest place any of us has ever seen, and it is hard to look, without wincing."

Thomas had forgotten about the Moon, of course. The stated position of NASA and many scientists was much the same, however, if less poetic. Premature overkill now replaced the earlier monster fantasies. As a result, support for further exploration dwindled and over twenty years went by before another scientific package was landed on Mars.

Mars Today: Rover Kisses the Rocks

Brian Cooper studied the terrain as he maneuvered the Rover, *Sojourner,* across the surface of the Ares Valles region of Mars. As the holder of the only Martian driver's license, signed in person by Governor Mike Lowry of the State of Washington, Cooper was fully qualified for the job. After all, as scientist Henry Moore of the Rover team put it, "Nobody has ever driven a car on Mars before."

The license was granted on grounds of analogy rather than law. An area in western Washington, called the Channeled Scablands, owes its unusual appearance to enormous floods that swept through the area at the end of the last ice age, 16,000 years ago. The deluge left many characteristic landforms spread over a large, gently rolling area. The diagnostic signs included low streamlined hills aligned with the flood direction, bars of transported debris, grooves eroded in rocks, channels, giant ripple marks, potholes, scoured-out lakes, and a great variety of boulders.

The appearance of Ares Valles, as viewed from above by the orbiters that accompanied the Viking landers, suggested that it had experienced a history much like that of the Channeled Scablands. Citing the presence of "deep, sinuous channels," "teardrop-shaped islands," and scour patterns, Michael Carr of the U.S. Geological Survey concluded that "the resemblance between the two is so strong, and there is so much other supportive evidence for water erosion, that a flood origin for most of the [Martian] outflow channels can hardly be doubted."

The Rover was transported to Mars as an integral part of the Pathfinder mission, the first craft to return to the Martian surface since Viking. The chosen landing site fell within an ellipse some 150 kilometers north of the mouth of a channel that had drained a highland area, Margaritifer Terra. The channel then flowed into the huge area called Chryse Planitia (the Plain of Gold). The *Viking 1* lander had settled in that area, farther from the channel mouth and 850 kilometers away from the Pathfinder site.

The floods that poured through Ares Valles and into the plain were of much greater magnitude, however, than any recorded on Earth. They were "almost 100 times as large as the largest known terrestrial floods," according to Carr. They may have involved the sudden eruption of water trapped beneath the frozen surface and released by an earthquake or impact. The Martian floods also differed from those in Washington State in that they occurred billions of years earlier.

At that distant time, the climate of Mars was believed to be more temperate than it is today, with liquid water on the surface. The reasons for this are not fully understood, but irregular changes in the tilt of the planet's axis may be involved. The amount of climate improvement at that time has also been a subject for argument. The *Scientific American* for November 1996 bore on its cover the words, "The red planet as water world: Mars had lakes, rivers and an ocean." The cover illustration showed barren orange mountainous terrain, but with a lake and a watery crater in the foreground. Others have suggested much more Antarctic conditions on Mars at that time, with the normal temperatures near −40°C (−40°F). Surface water would originate below the ground, rather than from rain, and would exist in liquid form on the surface for only a short time.

The Pathfinder lander, like *Viking 1*, observed a fairly level, rock-and-boulder-filled plain. The colors of the land, as displayed in NASA photographs, were a muted harmony of dull reds, oranges, grays, even greenish tints. Two low hills named Twin Peaks lay on the horizon with a pink-orange sky above. The scene appeared benign, if barren; it fell to the instruments to tell why unprotected humans would perish at the site. The temperature, for example, was at −53°C (−64°F). This was not due to an unusual cold spell but was normal for the site, whose temperature could reach −12°C (10°F) at noon, but as the planet rotated through its twenty-six-hour day, night inevitably came, and the temperature crashed to −73°C (−100°F). A very warmly dressed human could survive on the surface for much of that range but would find the atmosphere much more deadly. Its pressure is less than 1 percent of what we are used to on Earth, and 95 percent of that meager amount is our waste product, carbon dioxide. The remainder is nitrogen and argon, with almost no oxygen. A visitor without an appropriate space suit would quickly suffocate.

Brian Cooper was not inconvenienced by these conditions as he drove his Rover, which had been named after the nineteenth-century civil rights crusader, Sojourner Truth. I was able to interview him one morning, five days into the mission. I saw a friendly, relaxed, youngish man with his face framed by a full head of hair, a dark beard, and moustache.

He was comfortably dressed in a green sports shirt and khaki trousers. We were not on Mars at all, but in the press reception area, on the campuslike grounds of the Jet Propulsion Laboratory (JPL) in Pasadena, California. He would "drive" the Rover by remote control from a nearby comfortably air-conditioned room. He had already completed his real driving for that morning: an eight-minute trip from his home in nearby Altadena to JPL.

The first task of his routine was to put on 3D goggles and, using a Silicon Graphics workstation, examine the terrain in which the vehicle stood. The immediate goals would then be discussed with mission scientists, and they would select a target for the next move, plotting a day's travel with a joystick and a Rover icon. Over ten minutes would be needed for their instructions to travel the 190 million kilometers to the vehicle, and an equal amount of time for any return message. Further, the Rover could accept the instructions only while at rest and not when it was in motion. Obviously, a different driving strategy had to be devised than the ones normally used on a Los Angeles freeway.

When the team had selected the best path to the target on the screen, they would send commands to the Rover at the next opportunity for communication, using the nearby lander as an intermediary. Once the Sun had risen over the Mars landing site, *Sojourner* would proceed on its own across the Martian surface. Brian Cooper was unlikely to get a speeding ticket, as the Rover's top speed was about 60 centimeters per minute. With the help of its own computer, it would cope with difficulties and try to reach the goal.

The Rover was roughly the size of the coffee table that I keep in my living room and, like that table, had a flat platform on top. The resemblance ended there, as the Rover had six independent wheels in place of legs and could roll over small rocks rather easily. It got caught on larger ones on a couple of occasions, but the team was able to maneuver it free. If their efforts had failed, a tow truck would have been very long in arriving! Despite the mechanical nature of the Rover, I found myself empathizing with it as if it were a dog, as it crawled across the Martian surface. Perhaps our experience with the robot R2D2 in the *Star Wars* series has trained us to regard such machines as if they were our faithful servant.

In their initial specifications, NASA had planned to operate the Rover for at least seven Martian twenty-six-hour days, or "sols," but they hoped that it would function for thirty sols, the nominal life of the lander. In fact, both operated from July 4 to September 27, 1997, sol 83 of the mission. At that point, useful communication with the lander failed; the Rover presumably remained operable for a time. It undoubtedly sits there today, turned to the lander and patiently awaiting its next instruction.

The surface activities of the Pathfinder mission were an engineering success, and the landing itself even more so. A central purpose had been to show that an instrument package could be landed on Mars cheaply and safely. Cheap, of course, is a matter of perspective. The costs of the mission have been estimated at from $150 to $227 million. Its predecessor, the Viking project, had put two landers at widely separated sites on Mars in 1976, but at a cost of about $3 billion, in current dollars. Gross national product had gone up substantially in that period, but this was not accompanied by any desire to spend large sums on space missions. When the *Mars Observer* blew a hole in its fuel line while approaching the planet in 1993, nearly $1 billion was lost. This disaster had a sobering effect on the agency. The new NASA philosophy was "faster, cheaper, and better," and Pathfinder was an important part of the demonstration. But free lunches are no more common in space than on Earth. For a fraction of the cost, you get a fraction of the return.

Viking had dispatched duplicate orbiters and landers, each with a complex science package, while *Pathfinder* contained just a few selected instruments. The *Viking* landing sites were chosen after the spacecraft had settled into Mars orbit. The *Pathfinder* used a more audacious strategy when it landed on July 4, 1997. It flew directly in. A heat shield, parachute, and brief bursts from rockets brought the lander smoothly to within 15 meters of the surface, but it essentially dropped for the remainder of the distance. It hit the ground at 50 kilometers per hour, but ingenious airbags cushioned the shock. The lander bounced at least fifteen times, then rolled for ninety-two seconds before finally coming to a halt. It had four sides and could have landed on any, but the right orientation was needed for it to function. A procedure was available to flip it over, but it wasn't needed, as *Pathfinder* landed right side up. The lander unfolded and deployed three petal-shaped flaps. When the signal was received at JPL that all had gone well, the staff again cheered and hugged, as their predecessors had done for *Viking*.

The Pathfinder scientists had ample reason to be nervous. Despite the bounce capability, a direct hit on a large boulder would have penetrated the cushioning and wrecked the landing. During the simulations in the Channeled Scablands, JPL scientists had visited one particularly disturbing place called Monsters of Rock, "the most hazardous, rockiest site" in the area. In the discussion that evening, they felt "considerable discomfort with the size, roundness and percentage of boulders covering the surface at that site." As one participant put it, the "Monster Rock site is terrifying." The largest boulders were too large for the airbags to handle. Peace was attained when that site was declared atypical, and a more typical prototype was visited the next morning. In the words of the summary report, "This stop was the most important event of the field trip, because the engineers has grown very concerned about the safety of landing on a flood deposit, until they had an opportunity to see and discuss this site."

The Ares Valles site had its problems, but there was not a lot of choice about where to land because of the budget-conscious design of the mission. The need for maximum solar power generation and efficient communication with Earth constrained the latitude to the band 10 to 20 degrees north of the equator. Only sites at low altitude could be considered, because the parachute needed to operate over a significant distance to do its job. Pathfinder project scientist Matthew Golombek described the situation in advance: "There are only four tiny little sites on Mars where you can land."

Despite these limitations, the scientists had high expectations for the chosen area. It was hoped that the site would contain a "grab bag" of rocks, a large variety of different types washed down from the highlands by the floods. As Golombek put it, "If we're fortunate, we could characterize a huge part of Mars—how it formed, what the early environment was like— all from landing at a single point." But planetary geologist Larry Crumpler added, "It's hard to imagine a place as dusty as Mars is. Everything is covered with fine, flourlike dust." The danger existed that the instrument could end up determining the composition of a uniform coating of dust.

The principal instrument used by the Rover to study the rocks of Mars was an alpha-proton X-ray spectrophotometer. This device carried the radioactive element curium, which, when placed close to a target, bombarded it with alpha particles. The instrument could measure the radiation given off by the target and learn the identity and amounts of elements in it. In practice, the test involved maneuvering the backside of the Rover against the rock to "kiss" it, and gathering in the data that resulted.

After the landing, there was some initial concern as the cushioning appeared to be blocking the release of the Rover. But soon it was disentangled and *Sojourner* rolled onto the Martian surface. It spent the next months moving around within a few meters of the lander and sampling one rock after another, ones that bore irreverent names like Yogi, Barnacle Bill,

Lamb, and Scoobie Doo. Striking photographs and panoramas were produced by the lander camera showing the Rover against the background of Mars, and vice versa. A huge amount of data was released about the atmosphere, temperature, magnetism, dust cover, and weather. As one account summed up, "The scientists are like kids in a candy store."

In the end, however, it seemed that only one type of candy had been served. Richard A. Kerr reported in *Science* on January 9, 1998, "Pathfinder researchers had hoped that the landing site would reveal a grab bag of different rock types deposited by great ancient floods, and initial results supported that idea. But, at the fall meeting of the American Geophysical Union here last month, many Pathfinder researchers suggested that a single rock type lies behind the varied shapes, colors and textures that Pathfinder observed." Apparently they represented the same volcanic bedrock that had been treated differently, and they were covered with variable amounts of a sulfur-containing dust. The dust was widely distributed on the Martian surface, but its origin was a mystery. (The Pathfinder designers had not included a dust broom with the Rover.)

The mission hit its peak with the successful landing and in the days when the first color pictures were returned. Madeline Jacobs could write in *Chemical and Engineering News*, "I haven't felt this excited about space exploration since Neil Armstrong took his first step on the moon nearly 28 years ago." (The Viking mission was apparently forgotten.) The unprecedented number of "hits" on the Pathfinder Internet sites reinforced her conclusion. *Time* magazine and *Newsweek* both put photos of the Rover on Mars on their front covers. But by late the next week, when I visited JPL, the excitement was already cooling. Many vacant seats had appeared at the press conference.

On July 9, Pathfinder was still front page news in the local paper, the *Pasadena Star-News*. But the headline said, "Photos from Mars: Is the thrill already gone?" The inside story, written by Matt Crinson of the Associated Press, carried the following comments:

> The planet looks just as it did when the Viking missions landed there in 1976—rocky, barren and red.
>
> Some Americans are beginning to ask: What's the big deal?
>
> To the layman, Pathfinder has made excellent television, with teary-eyed mission controllers jumping up and down, excited scientists spouting technical terms and dramatic martian landscapes flashed on the screen.

But the mission's scientific significance is much less exciting. The great variety of rocks that has geologists agog looks pretty uniform to the average person.

Viewing the Desolation

I had wanted to share the Pathfinder experience in some way, and sense the terrain. As I could hardly visit Ares Valles, I did the next best thing. In July 1997, as the Mars *Pathfinder* Rover was being driven across the surface of Ares Valles, my wife Sandy and I made an excursion to the Channeled Scablands. We followed in the steps of the group of more than sixty Pathfinder scientists and public school educators, who had conducted their workshop there two years earlier and tested a Rover prototype.

We drove north from Moses Lake, Washington, through a region that rolled gently and was largely unpopulated and somewhat desolate. Low ridges and small lakes were visible as we headed north. The temperature was warm but comfortable. We stopped near a collection of rocks and boulders that was considered representative of those of the Pathfinder site. They were scattered among low rises and gullies. I remembered the elaborate strategies that were needed to have *Sojourner* sample a Martian rock, and how the analysis was confounded by the dust cover. In the Scablands I found it quite easy to pick up any rock that I fancied, dust it off, and examine it as closely as I chose. Sending humans to Mars and returning them safely will be much more difficult and expensive than an excursion to central Washington State, but the advantages of an on-site human presence are also overwhelming.

The Scablands were relatively unpopulated, but roads and power lines kept us aware that we were on Earth and not on Mars. Even when I focused my attention down on the rock area, the presence of dry yellow-green grassy weeds and dandelion-like flowers between the rocks reminded me of where I was. I did not wander too far from the road in my exploration, as the earlier group had discovered rattlesnakes in the area. This health hazard was tiny, however, compared to those humans would face at the Ares Valles landing site.

At that site, the story lay not so much in what was seen at the site as in what was not seen, just as it had been for Apollo. The desolate desert photographed by *Pathfinder* was much like that observed by the Viking landers. Before Viking, Carl Sagan had written in the *New York Times*, "There is no reason to exclude from Mars organisms ranging in size from ants to polar bears." But afterward, he lamented, "There was not a hint of life—no bushes, no trees, no cactus, no giraffes, antelopes, or rabbits, no burrows, footprints, or spoor. . . . There was not a single recognizable funny-looking thing, no obvious sign of thermodynamic disequilibrium." The same could be said for the Pathfinder site, and remembering the Channeled Scablands, I might add weeds and rattlesnakes to the list of items not seen on Mars.

Similar considerations inspired Roger Rosenblatt to write an article for *Time* magazine on July 14, 1997. In it, a subheadline announced: "Pictures sent back from Mars prove just how alone in the universe we are." The article continued: "Out there is Mars with its wasted territory. . . . If life existed anywhere else under the sun, it should have been there. . . . Errands into space lift us out of ourselves and return us to ourselves. They tell us that we are alone in the universe and how terrible and wonderful an idea that is."

It is tempting to observe a barren wilderness and draw sweeping conclusions about the universe, particularly if our need for an all-encompassing story tilts us in that direction. But before we take that journey, we must give nature more of a chance to tell us what is going on. The following warning signs can keep us alert.

1. Three locations do not define a planet. The two Viking sites and the one chosen by Pathfinder had similar appearances. This should not be surprising, as comparable considerations of lander safety were involved in the selection of each of them. Desolate areas with the same appearance can also be found on Earth. One location in Death Valley National Park has been named Mars Hill by the U.S. Geological Survey because the terrain on this hilltop is similar to those of the Viking sites on Mars. It was used for some earlier Rover tests. Many other such barren-looking desert sites can be seen on this planet. Yet very different surroundings may exist only a short distance away.

After Sandy and I had visited the Channeled Scablands, we were able to drive for just a few hours and arrive at Mt. Rainier, a spectacular snow-covered volcano. Similarly, a day's drive from Death Valley can bring you to the edge of the Grand Canyon. There are no roads on Mars, but even more impressive wonders lie within 2,000 kilometers of the Chryse Platinia plateau. The volcanoes of the Tharsis region, for example, dwarf Mt. Rainier and even Mt. Everest in height. The enormous canyon of

Valles Marineris, which lies south of Chryse Planitia, extends for a distance comparable to the width of the continental United States. By contrast, much of the northern hemisphere of Mars is a low-lying remarkably flat plain, which extends for thousands of kilometers. Geophysicist Maria Zuber of MIT commented that it is "the flattest surface in the solar system for which we have data." At the poles of Mars lie substantial icecaps, which vary in size with the season. The one at the North Pole, like that on Earth, is composed of water ice. But the Martian South Polar cap has a core of dry ice, solid carbon dioxide. All of these environments await our exploration, and all of them need to be sampled before we can draw any sweeping conclusions about life on Mars.

During the Pathfinder mission, the Rover spent its time entirely in the vicinity of the lander. The *New York Times* had reported on August 22, 1997, that ultimately, the Rover was to head for a low hill a few hundred feet to the north "to look over the edge and see what a whole new vista looks like." This event never took place before contact was lost. Most likely, we would only have seen more of the same. But if we embrace the idea of exploration, and do enough of it, we may occasionally be surprised.

Physicist Freeman Dyson has suggested, for example, that warm-blooded plants could survive on the Martian surface within greenhouses grown of their own flesh. These structures would allow light to enter inside for photosynthesis while conserving heat and oxygen. The plants would use deep roots to draw water and nutrients from the underground. Dyson concedes that the discovery of such plants is only a "remote possibility." But we cannot test such ideas unless we look. If we choose to send robots, and never ourselves, and program them in advance to detect only things that we expect, our results will be very boring.

2. Photographs do not tell the whole story. For example, the locations on Earth that are considered most Mars-like are certain frigid and dry (by Earth standards) valleys in Antarctica. Tests on the soil from such regions show them to be almost free of organic compounds, as was found at the Viking sites on Mars. Yet hardy lichen and bacteria were found living within the rocks. They function at a depth that protects them from freezing yet allows them access to sunlight. They are inactive for much of the year but rouse themselves when the temperatures warm up.

Freeman Dyson speculated that his warm-blooded plants might (or might not) be readily seen in photographs. He felt that we would do better by seeking their heat radiation at night, or by detecting the oxygen leaked by them. When we search for unknown life forms, we will be wise to keep our senses alert and a host of detectors at our service. What we find may surprise our expectations.

3. Remember Gulliver. "Gulliver" was the name given by Dr. Gilbert Levin to the earliest homemade versions of his own life-detection apparatus. I find this name easier to keep in mind than the more official title of Labeled Release, or LR.

Levin is the president of Biospherics, Inc., an environmental consulting company located in suburban Maryland. He has also had a long-term personal involvement with the search for life on Mars. It appeared to him that he stood at a disadvantage when proposals were requested by NASA for instruments for the Viking flights to Mars. He was competing with scientists located within NASA and others with formidable academic credentials. In the end, however, he was among the three winners, along with Professor Norman Horowitz of the California Institute of Technology and Vance Oyama of NASA. Each of the chosen experiments attempted, in a different way, to demonstrate that life was present on Mars. Each was based on a particular assumption, and none was intended to be comprehensive. In the pre-Viking words of astronomer Carl Sagan and Nobel Laureate bacteriologist Joshua Lederberg, "Negative results would exclude an important subset, but only a subset of the possible classes of Martian organisms."

The design of Levin's experiment was uncomplicated. A small drop of water containing five simple organic substances was placed in the center of a soil sample. As liquid water cannot exist on the surface of Mars, due to the freezing temperatures and low pressure, the test was carried out within the lander using Earth-like conditions.

The procedure had been tested exhaustively on our own planet. In all tests in which the sample contained Earth microbes, the microbes regarded the organic substances as food and happily consumed them. They converted their meal to carbon dioxide, which was released into the air above the solution. This event could be followed with great sensitivity, as the food was "labeled"; the carbon within it was radioactive. The release of radioactivity into the air above the water droplets announced the presence of life, hence the name "Labeled Release." If no label were released, the test was negative, though of course life forms that perished under the Earth-like conditions or had no appetite for the selected meal might still have been present.

Levin has summarized the results of tests on Earth: "Thousands of soils and microbial cultures of a wide variety of species were tested. No sample with demonstrated living organisms failed to respond."

Samples were taken from the dry and frigid valleys of Antarctica, which have extremely low microbial populations. They tested as positive but gave a low response. On the other hand, when samples were preheated at 160°C (320°F) to kill microorganisms, no radioactivity was released. A sample from the Moon also gave a negative result, without any heating, as did one from a freshly erupted volcanic island off Greenland.

There was some concern in advance that Martian organisms might be killed when exposed to liquid water at earthly temperatures. If they had adapted to the cold, dry conditions of Mars, this treatment might prove lethal, just as the oven heat was fatal to our own organisms. A false-negative would then be obtained. But, in fact, the samples tested at both Viking sites gave positive results, as defined by the standards published in advance of the mission. Gas was released, and this response was abolished when the samples were sterilized before the test.

Some anomalies were observed, however. The amount of radioactive gas that was released suggested that only a portion of the food had been consumed. When a fresh round of food was added, no new gas was released. The overall result still scored as positive and may have indicated that the organisms died during the procedure.

Because of the importance of the positive result, an unanticipated test was improvised at the Viking site. Some Viking scientists felt it possible that chemicals that did not normally occur in samples from Earth or the Moon were present in the Martian soil and responsible for the results. The whole point of the sterilization procedure, of course, had been to distinguish life from chemicals that imitate a living response. Bacteria are destroyed at 160°C (320°F), but the most plausible candidate chemicals of this type would not be affected by such heat. To exclude the possibility that some unexpected heat-sensitive chemical was present and responsible for the result, it was proposed that the Viking soil samples be preheated at a much gentler 50°C (122°F).

The Viking team improvised a method in which the sample was preheated to 50°C rather than 160°C. They judged that this treatment was unlikely to affect nonliving chemical processes, but Martian sensitive organisms might be vulnerable, and the labeled release might be diminished or abolished. In fact, the procedure did destroy much of the ability of the sample to release carbon dioxide.

Harold Klein, the NASA biologist who was in charge of the overall life-detection package, later wrote, "In many respects, the LR (labeled release) data was entirely consistent with a biological interpretation. Indeed, if the information from other experiments aboard the *Viking* lander had not been available, this set of data would almost certainly have been interpreted as presumptive evidence for biology."

Of course, the various Viking experiments were based on different assumptions about Martian life. The package designers intended that each result be considered on its own. In the pre-Viking words of Richard Young of NASA, "It was felt that any one technique for life detection was unlikely to succeed, and therefore, a combination of techniques was required." In other words, it would not matter if the first two pitches were strikes, provided that the third was hit for a home run.

The baseball analogy fails at this point, however. In practice, the other two Viking life-detection tests were neither clearly positive nor clearly negative. Professor Horowitz's experiment gave marginally positive results, but not consistently so. The positive results should have been abolished by heat, if they were due to life, but this did not happen. Was some unforeseen chemistry at work here? Or was another strain of Martian bug responsible for the results? I feel it best not to draw any conclusion at all from such data.

Vance Oyama's experiment also gave an unexpected result. He planned to administer a complex nutrient mixture and actually identify the gases that were set free. But when water alone was added to Martian soil as a control, oxygen gas was released. Martian soil was then different from any other one that had been tested, earthly or lunar, with or without living organisms. This result said little about the presence or absence of life but made the remainder of the data difficult to interpret.

The other two life-detection experiments, then, told us little about the presence or absence of life. Apart from the psychological effect produced by the desolate appearance of the sites, the principal source of the negative inferences concerning Martian life was the gas chromatograph-mass spectrometer experiment. Life on Earth is almost always accompanied by a significant amount of organic debris in the soil, but the Martian soil had little or none. One explanation that appeals to me is that Levin's experiment detected dormant spores on the surface. They had been blown by the wind from some environment that was richer in organisms and debris.

The Labeled Release experiment clearly stands as evidence for present life on Mars, just as the ALH84001 data serve as evidence of life in the past. Once again, though, we run into the maxim that extraordinary claims deserve extraordinary proof. Such limited sets of data are not strong enough to support the weight of the conclusion that extraterrestrial life exists and that a Life Principle may be at work in the universe. But they should offer encouragement, a strong invitation for scientists to collect more data until the issue is settled. In the case of the Viking results, surprisingly, the exact opposite happened.

An exhibit on Viking in the Smithsonian Institute in Washington bears the statement, "The Viking Lander Biology Experiments searched but did not find any evidence of life on Mars." In a display at the American Museum of Natural History in New York, I saw a similar claim: "No evidence was found." Similar summaries can be found in many sources in the media; ultimately they can be traced back to scientists in NASA. The limited excerpts that I have presented just display a confusion between evidence and proof. My dictionary says that *proof* is "evidence sufficient to establish a thing as true, or produce belief in its truth." *Evidence*, on the other hand, represents less. It is information that argues for one interpretation of events (for example, eyewitness testimony of a murder). But evidence can be misleading or mistaken. In this light, the LR experiment found evidence for life on Mars, but not proof.

Much more than semantic confusion lies behind the Viking interpretations that reached the media. A Sour Lemon stampede followed that mission. The statement by Professor Horowitz at the start of this chapter provides a fine example, but many more are available. The reasons for this are less clear. Barry DiGregorio, who has written a history of the LR experiment and its aftermath, has offered the following diagnosis: "After Viking no one wanted to be associated with anyone who claimed there was life on Mars, for fear of being branded a Percival Lowell. Everyone wanted to play it safe and ride on the politically correct 'dead Mars Bandwagon.' "

In that same book, Gilbert Levin put forth his own opinion:

> I think that once NASA announced that the LR had produced "no evidence" for life, the scientists in the agency, outside scientists supported by the agency, and those looking to the agency for future funding took their cue. Consciously or even without overt intent, they coalesced behind the official opinion—even to the extent of using that patently inappropriate phrase "no evidence" to describe the LR's findings. The unlikely alternative is to believe that a large number of scientists do not know the meaning of the word "evidence"!

Levin went on to describe his own post-Viking experiences:

In 1986, in a talk at the National Academy of Sciences celebrating the tenth anniversary of Viking, I said that "more probably than not" the LR had detected microbial life in the soil of Mars. This produced near-pandemonium among the scientific audience. At the reception, which followed the talks, prominent Viking scientists accused me of having disgraced myself and science.

Levin subsequently tried to have an updated version of Gulliver included in future Mars landing missions. His ideas drew support in Russia but received short shrift in this country. Certainly none was considered for Pathfinder, nor is any on the agenda for presently scheduled U.S. missions.

Yet in July 1997, as the *Pathfinder* Rover was inching across the surface of Ares Valles, I saw Levin appear as a featured speaker at a convention of Optical Engineers in San Diego. A special symposium had been arranged on the detection of extraterrestrial microbes, with key personnel from NASA in attendance. When Levin mounted the podium to present his data once again, he was greeted with thunderous applause. What had happened?

A Shift in the Tide

The Sour Lemon School point of view that dominated the discussions of life on Mars after Viking not only quenched public enthusiasm but also delayed future exploration of that planet. A follow-up mission to Mars in 1984 was planned by a working group chaired by geologist Thomas Mutch. The group's report stated, "The odds given by various scientists for the existence of life on Mars today vary considerably, but they are usually small. It is therefore reasonable to assign higher priority in Mars '84 to the exploration of geophysical and geochemical questions."

The instrumentation that the group proposed had a considerable potential for chemical analysis and might even have shed further light on Gil Levin's results, but the group's stated science objectives emphasized the study of the minerals, landforms, atmosphere, and magnetic field of Mars. No funds were voted for this mission by Congress.

Nature and fate then introduced their own sour notes. New efforts were made to explore Mars and its satellites from orbit, and all of them failed. Two satellites were sent by the Soviet Union in 1988 to examine the

Martian moon, Phobos. One of them was lost because of a ground control error, and the other broke off contact when within sight of the Moon. The expensive Mars Observer mission of 1993 was intended to improve the maps prepared by Viking, but it apparently blew a hole in a fuel line and went out of control. More recently, the Russians had organized an ambitious *Mars 96* lander. This vehicle carried a number of U.S.- and European-designed experiments and a list of the names all of the members of the U.S. Planetary Society (including my own). Alas, a booster failed and the experiments and names ended up in the South Pacific rather than on the cold deserts of Mars.

When the Pathfinder engineers successfully landed their instruments on the surface of Mars, they ended a twenty-year drought in the closeup study of the planet. During that period, other scientists have also struggled to turn another tide, one that ran against the biological investigation of Mars. Chris McKay, Penelope Boston, and other graduate students had been thrilled by the Viking landings and their search for life. In the 1980s they established an unofficial network to promote the further exploration of Mars, particularly by humans. Eventually their group gained the name the Mars Underground. A newsletter now records their progress. The *Mars Underground News* arrives on my desk four times a year with news of their quest, and in the past few years the news has grown more promising.

With the passing of time, this group and other scientists in NASA came to realize that early Mars may have had an environment similar to the early Earth. If that were the case, then life may have started on Mars when it did on Earth. Even if that life died out when the climate changed, it may have left behind some record that we can detect. In the 1990s, the idea of a search, at least for signs of extinct life, had entered NASA workshops once again. Finally, the 1996 report of possible fossils in Antarctic meteorite ALH84001 gave new respectability to the quest. The missions for the next several years, like Pathfinder, had been locked in place with their geologically based agendas. But now some limited exploration for life could be introduced again.

The Great Life Hunt

If we make the favorable assumption that life began on Mars 4 billion years ago, how could we find decisive proof of that on the very different Mars of today? Chris McKay, now at NASA, has worked with a presumed history of life on that planet.

He starts with the era when life began and liquid water existed on the surface. Gradually, as pressures and temperatures fell, the surface of the planet became less habitable, but ice-covered lakes, warmed by heat from the interior, could have provided a refuge for Martian microbes. Such habitats can be found on Earth today in dry, cold Antarctic valleys. On Mars, environments of this type may have survived for up to a billion years. During that time, sediments deposited at the bottom of the lakes may have provided a fossil record of the organisms.

When the lakes finally froze solid, or lost their water in other ways, only sites within the rocks would remain. In the final phase from 1.5 billion years ago to the present, those last refuges also lost water, falling victim to the cold and loss of atmosphere. In Chris McKay's scenario, life then disappeared from the Martian surface, but not necessarily from Mars.

On this planet, microbial communities have been found thriving far below the ground. They have been recovered from porous rock formations as far as 2.8 kilometers down, at temperatures up to 75°C (167°F). They live on chemical energy, combining buried organic matter with minerals from the Earth. Physicist Thomas Gold has argued that such microbes need not depend on debris that comes down from the surface for their food. He believes that the main components of petroleum may be formed deep within the Earth and well up naturally, providing a reliable nutrient supply for underground life. He suggests that "this new biosphere would be comparable in mass and volume to all the surface life we know." Further, such a community, which he calls "The Deep, Hot Biosphere," could exist on any planet large enough to retain appreciable internal heat. In our present state of ignorance, we cannot rule out the possibility of a biosphere of this type, a true "Martian Underground," on Mars.

The advocates of the hot and deep biosphere are complemented by those who base their hopes on a frozen one. David Gilichinsky and his colleagues at the Puschino Institute in Russia have recovered microorganisms that survived for 3.5 million years after storage in permanently frozen Siberian subsoil (permafrost) at −10°C (14°F). Chris McKay has suggested that the southern polar regions of Mars may have preserved the frozen remains of Martian microbes at much lower temperatures for vastly longer times. He feels that they would have been killed by radiation during a very long storage, but their preserved remains would tell us about their biochemistry. For my own part, I would prefer to carry out the search first and learn about their survival after the discovery has been made. If their chemistry were truly different from ours, then their vulnerability to radiation could be much less than our own.

The next vehicle to land on Mars, the Mars *Surveyor 1998* lander, will actually set down in the southern polar region. Unfortunately, it was designed at the time when geological questions were in the foreground and has "Volatiles and Climate History" for its theme. Despite the date in its name, it will be launched in early January 1999 and arrive on Mars that December.

Cameras will record both its descent (it will use a heat shield, parachute, and rockets) and the environment in which it arrives. Its surroundings should be quite photogenic, as it is targeted for the edge of the carbon dioxide icecap, in a region described by Jet Propulsion Laboratory as "strange, layered terrain." In keeping with its theme, the lander will study the atmosphere, dust, and weather and dig into the soil with a robot arm to look for evidence of water ice below the surface.

Although the lander has no wandering Rover with it, it will launch two separate penetrators as it descends, which will slam into the surface at 200 meters per second (450 miles per hour). They will also make geological measurements, including the amount of ice below the soil.

Neither the lander nor the penetrators have the capacity to test for life directly, though if a polar bear should wander within camera range of the lander, it would be noticed, and a miniature microphone would record its grunts. But more ambitious searches have been scheduled for the early part of the next century.

The Martian Olympic Games

A variety of locations on Mars, far greater than the number that seek the Olympic Games here, are competing for an ultimate honor: selection for a 2005 Sample Return Mission. Material from the chosen site will arrive back here in 2008 but will be more likely greeted by a strict quarantine than with a ticker-tape parade.

Additional robotic exploration of the Martian surface will precede the final selection, provided that NASA is given the necessary resources. Budget cuts present as real a hazard to future missions as the failure of rocket components and the presence of boulders at the landing site. In the springs of 2001 and 2003, new landers will be dispatched to separate sites on Mars. The 2001 lander may carry the *Marie Curie* Rover, a sister of *Sojourner*, or, if the necessary funds are not appropriated, no rover at all. The 2003 mission will carry an improved rover, one that will have the capacity to travel up to 100 kilometers over a year, examine mineral textures under a microscope, and collect samples. The plans for the 2005 Sample Return mission had initially included a rendezvous with a rover from the 2001 or 2003 mission, to retrieve samples that had already been collected and return them to Earth. By mid-1998 these plans had been altered. The 2005 lander would collect its own sample and return it, with the earlier rovers serving only for surface exploration.

One avowed purpose of this sequence of missions is to collect evidence of life on Mars. But questions of strategy remain undecided: Where should they land, and what should they seek? A number of landing sites are already being studied for the 2003 Rover, which has been named *Athena*, after the Greek goddess of wisdom. Many of the same technical limitations still apply that held for *Sojourner*, the *Pathfinder* Rover. The landing site must fall within a band of about 15° around the equator if it is to generate enough solar power. Polar sites, with their possible frozen microbes, are excluded for this reason. No deep excavation in search of an underground biosphere can be undertaken in the near future. According to paleontologist Jack Farmer of NASA, "access to such environments will likely require drilling to depths of several kilometers." But samples from below may be waiting for us on the surface, if we can locate on Mars the remnants of a feature that we know very well on Earth.

At Yellowstone Park in the American West, and in a number of other places on Earth, hot springs and geysers bubble or spurt hot water onto the surface, bringing underground materials with them. The springs also furnish a favorite growing place for microbes. The mineral deposits formed by the springs provide samples of what lies below as well as microfossils of unlucky bacteria trapped during mineral formation. The residues surrounding the site of an ancient Martian hot spring would be prime targets for a microfossil search. But before we can dig we must find such a site, as none have been identified on Mars so far. Some more thorough reconnaissance from above will be needed.

A series of *Surveyor* orbiters will be mapping the surface in detail in the next several years, measuring the height, ruggedness, temperature, mineral content, and dust cover of the Martian landscape. The first, *Mars*

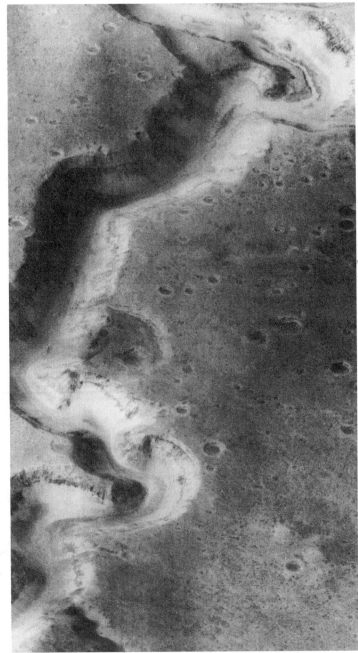

The *Mars Surveyor 96* orbiter (Mars Global Surveyor spacecraft) obtained this photograph of the 2.5-kilometer-wide Nanedi Vallis canyon in January 1998. Rocky outcroppings are visible along the upper canyon walls; weathered debris lies along the lower canyon slopes and on the canyon floor. The features are ambiguous. Some suggest that the canyon was carved by continuous flow of liquid, while others suggest formation through collapse. (NASA/JPL/Caltech)

Surveyor 96, arrived safely but encountered technical delays in getting into its final orbit, which has been delayed until 1999. It has returned useful photographs while making its maneuvers, and more can be expected when it reaches its proper position. *Mars Surveyor 98* will join it near the end of 1999, and additional orbiters will add to the party at two-year intervals after that.

The data from these orbiters will provide crucial input for the site selection committee. Some luck will be needed if they are to locate the remnants of ancient hot springs. Jack Farmer of NASA has pointed out that such targets on Earth usually occupy an area of a few square kilometers, which is below the resolution of the orbiting instrument best equipped to detect them.

In the absence of such a site, Mars scientists have suggested two other kinds of locations in which sediments may have been laid down underwater, creating microfossils. The dry beds of ancient lakes, many of which formed in craters, have attracted a following. They offer not only the possibility of preserved biological remnants but also safe and practical landing targets. Another group of sites features the side canyons of the huge Valles Marineris, whose layered appearance suggests the presence of water-deposited sediments. Our previous experience on this planet has indicated that they might be promising locations for seeking microfossils. But much greater landing precision would be required there.

One way or another, a site will be selected, and the 2003 *Mars Lander* will descend there softly on its three legs. The *Athena* Rover will emerge and set out across the landscape with much more energy than did its predecessor, *Sojourner.* The vehicle will move up to 300 meters per day, and at least 10 kilometers overall during a one-year lifetime, scanning the terrain for promising mineral environments. It will approach suitable rock formations, "kiss" them in the style of *Sojourner,* and examine them with a microscope as well. When *Athena* meets a worthy candidate, it will drill into it using a Mini-Corer and transfer the sample into one of ninety-one core collection bins at its front side. If the rendezvous concept for Martian sample return should be restored, then these cores, together with some soil samples, will have earned a trip back to Earth.

In a more likely alternative, a roving vehicle from the 2005 Sample Return mission will collect its own specimens. The collector will then blast off, rendezvous with an orbiter, and begin the long return trip to Earth. For a cost of $500 million, 0.5 kilogram (1.1 pounds) of Martian rock and soil will arrive back on Earth in 2008. The reception planned for this package in a Mars Receiving Laboratory has been described as "worthy of the Ebola virus." This elaborate quarantine will protect us from the unlikely possibility that Martian organisms can infect us and also guard the sample from earthly contamination. After a suitable period has passed, the samples will be distributed to Earth laboratories for intensive study. What should we expect to learn?

Only the actual events will inform us. But I am made wary by our experience with the meteorite ALH84001. In a sense, we have already had a sample return from Mars. The scientists who studied that meteorite were able to interpret its contents in a very positive way, suggesting that life may have been present on early Mars. Yet the very nature of the evidence they presented made it impossible to draw strong conclusions from it. Other groups have disputed the evidence, item by item, but without being able to invalidate it. The entire question remains suspended in a state of limbo, joining Gilbert Levin's disputed Viking results.

Perhaps a sample return will produce the needed "smoking gun." But the circumstances surrounding it are not encouraging. John Kerridge of the University of California, San Diego, one of the scientists involved in the search, has called for a Rover equipped to do sophisticated analyses. In particular, it would search for samples containing organic materials and analyze them at the site. Any sample return would be delayed until a promising candidate had been found. But the 2003 *Athena* Rover and its successor in 2005 may not be so equipped. We must be prepared for the possibility that a returned sample will produce ambiguous results, or test negative, but come from a site selected after superficial examination.

If life has survived on Mars, it may be elusive. Bruce Jakosky, of the University of Colorado, has calculated that energy may be the limiting factor, with the Martian biomass less than one-millionth of the one on Earth. He suggests that volcano-heated sites may be among the best places to look, and the sites selected for landing safety the worst. If that should be true, then an extended series of investigations by better-equipped Rovers and multiple sample return missions from many areas of Mars may be needed. An extended, if remotely controlled, siege of Mars appears to be the current plan and cannot be faulted scientifically. Eventually, a mobile enough Rover will reach the right area. If life or fossils exist on that planet, they have been there for billions of years and can wait a few more to be discovered. But the situation on Earth is not as stable.

The grave of the Apollo program stands as a reminder that the public has little patience for extended enterprises where only dry data about geological formations, seismology, and the like are reported. Surface photographs of other planets also lose their freshness after repeated exposures. In the 1990s, the United States moved through a period of uninterrupted prosperity and relative international stability. Yet in that time NASA abandoned the ambitious and grand planetary surveys of the previous decades and settled for a series of much more constrained and less expensive projects, including robotic probes of Mars. If this is the harvest of good times, what will the next recession produce?

In addition, the record of the past tells us that a good fraction of the missions sent out will fail to arrive. Because of the political and engineering hazards, I believe that each lander should be treated as if it might be the last one, and include at least one direct life-detection experiment, in the style of Gulliver. At present it seems more likely that some future Russian or European mission will attempt direct life detection than one sponsored by the United States. In 1997, the European Space Agency approved the launch in 2003 of a mission called Mars Express. The plans may include a lander, equipped with instruments that can search for life.

But I suspect that this battle is one that, in the end, we will have to settle in person. In this spirit, I have created a vision of coming events.

Mars Future: The City in the Crater

June 18, 20-something: Janice Chen looked up from the controls of her Rover excavator and paused to stare again at the familiar terraced walls of Gusev Crater. As she enjoyed the view, her vehicle carried her slowly across the almost flat floor of the crater on its six tractor-tread wheels. The Rover's computers handled the burden of selecting a path, automatically avoiding the occasional deep ruts and large boulders and the remnants of the many deliveries of supplies and equipment that had been bounced, Pathfinder style, into the crater in the years preceding their mission. Janice needed only to scan the terrain for any subtle driving hazard that might elude the computer or an interesting feature that might motivate her to stop.

As she watched, the Rover changed its direction to swerve around a forest of tufa towers that jutted up from the crater floor. These fingerlike structures made of carbonate minerals were relics of the time when a lake had filled the amphitheater, billions of years ago. She had examined a similar group at Mono Lake in California, as part of the crash course in micropaleontology that was prescribed after her selection for this mission. The California structures had held a harvest of microfossils, remnants of the bacteria that had been trapped by the sudden deposition of calcium carbonate from the lake waters. After some practice, she had learned to prepare the necessary thin sections for microscopic examination and to recognize the more obvious artifacts that might pose as microfossils. Janice had to learn many skills in a hurry, to supplement the abundant qualifications that had earned her a place on the first Mars expedition.

Her skill as a physician had been central to her selection, together with her deep understanding of nutrition and waste management. She also loved the outdoors and had some prior knowledge of geology. In her role as leader of many wilderness backpacking expeditions, she had gained the psychological skills needed to keep peace in a small group sharing each other's company in an isolated setting. Finally, she had the physical qualifications needed, rugged stamina and agility packed into a body that, at most, registered 46 kilograms (102 pounds) on the scales used back on Earth.

When the budget-conscious mission planners had recognized that the payload to be lifted off of Mars for the return trip was a critical cost factor, the number of astronauts was cut back to four, and then it was recognized that four women could bring the same skills as four men and weigh a lot less.

At the time of the *Apollo 11* Moon mission, Vice President Spiro Agnew had commented, "I don't think we'd be out of line in saying we are going to put a man on Mars by the end of the century." He had not only gotten the century wrong but the man part as well. The decision to rely on an all-female crew had boosted Janice's chances considerably, but she was also helped by one less obvious qualification. She was willing to separate from her friends and family for years, because she had nursed a secret ambition: she wanted to find life on Mars.

As a child she had read science fiction avidly, particularly the older stories set on Mars. She had especially loved Ray Bradbury's *Martian Chronicles,* with their tales of the intelligent race that perished after we arrived. The Martians had "gold coin eyes" and rode insectlike six-legged machines with red and green lights that glittered like jewels. They lived in "white chess cities" with crystal windows and fragile towers, beside the "green wine canals."

Janice understood that this had been written as fantasy, appropriate for the time just after World War II. The Martian natives, if any existed, were more likely to resemble what the *New York Times* had once termed "shower scum." But just the same, Janice felt that there would be something wonderful about such a discovery, and now she was the person with the best opportunity to make it. She alone, of the group of four who had made the trip to Mars, had the necessary background in biology. Any search that she made, however, would be an unauthorized addition to her assigned duties.

The official position of NASA, echoed by the media, was that Mars was lifeless. Only fossils, remnants of the early era when Mars was wet, might be expected. If the possibility of current life was raised again, then expensive sterilization measures would be required, and the cost driven well beyond reach. Hence no life-detection tests could be included on the mission.

This position had remained firmly in place since the early days of the twenty-first century, when a collection of samples had been returned from Mars. Only small amounts of organic material had been detected. Traces of left-handed amino acids had been found in a few of the samples, but these were regarded as remnants of meteorite infall, or perhaps signs of much earlier life. The microfossil question remained open, as one sample had contained minerals and microscopic shapes much like those in the ALH84001 meteorite. Additional robot explorations and sample returns had been planned, but the program was abruptly terminated by Congress during a recession.

With this background, Janice felt lucky to be on Mars at all. Interest in space had reawakened, however, when an asteroid came close to Earth in 2028. New concerns had also come up when a global cooling replaced the expected climate warming caused by carbon dioxide emissions. Was this new effect caused by human activities, or was the entire solar system perhaps entering a cloud of interstellar dust? Such a cloud, by screening out heat from the Sun, could start a new Ice Age. The study of climate changes on other planets had suddenly taken on great importance. A human mission to Mars was put together as quickly as possible with private support from the Mars Society and other groups filling in the gap in government appropriations.

Janice fully supported the announced mission objectives, but she brought along her own, unadvertised agenda. She had constructed a mini-Gulliver machine, using bits of hardware left over from the construction of their base and a few key items that she had taken along among her personal effects. Although the returned Martian sample had given only negative results with Gilbert Levin's procedure, Janice felt that Martian soils would behave differently when examined on their own planet. She had been right in this expectation, but the positive Gulliver results had been the only encouraging sign so far.

A sudden change in direction by the Rover brought her out of her reverie. The vehicle had completed its detour around the tufa towers and resumed its original direction. At the start of the mission, these towers had been a citadel for her hopes; now they were simply an obstacle. Under the microscope, they had appeared quite different from their counterparts on Earth. The carbonate layers here held only irregular fragments of a sulfur-containing mineral, interspersed with bubblelike enclosures that, by Earth standards, were too small to represent cells. Janice had done no better when she sampled a variety of features near the crater: the sediments deposited by Ma'adim Vallis, the river that had once emptied into Gusev, the cone-shaped hills that rose in places from the ancient lake floor, some terraces surrounding a smaller crater within Gusev, and the extensive streaks of wind-blown darker dust and debris that marked the landscape. Only Janice's positive Gulliver results kept her going.

The response was always weakly positive but stronger in some places than others. When Janice marked the pattern on her map, it pointed in a definite direction, and the Rover was following that trail.

The vehicle finally pulled up at the site of the strongest response, a tiny crevasse that emitted small amounts of warmer gases that included traces of water vapor. The emissions suggested that hot springs existed below the site; they may have been generated by the impact that created the crater. Janice had reported her discovery, omitting the Gulliver tests that had led her to it, and she had received permission to excavate at the site.

She now checked the air supply on her spacesuit and found it adequate for many hours; the suit, her equipment, and the Rover were all in fine working condition. Only a few clouds appeared in the pinkish sky, and the temperature was relatively warm for the location. In all, it was a fine morning on Mars. She could go ahead with the experiment.

She climbed down to the bottom of the depression that she had bulldozed and into a small pit that she had shoveled out herself. A small tent at its bottom protected a video terminal, which drew its signal from an apparatus that she had inserted into the ground. The terminal provided a magnified view of a small illuminated chamber several meters underground.

The apparatus also allowed her to insert and manipulate materials within the chamber.

She began her routine by reading the temperature and humidity in the underground enclosure. Both were higher than on the surface, as usual. She could proceed with her experiment. At her command, a small quantity of powder was placed on the floor of the enclosure. It contained the left-handed form of an amino acid, one that had been found in traces on the Martian surface and in the samples returned to Earth. This form was, of course, the one preferred by life on Earth. She then waited and watched to see whether any creatures would be attracted by the bait.

Hours later, she was about to abandon her effort. Nothing whatsoever had taken place within the chamber. Suddenly a new thought hit her. The food that she left over at dinnertime when she was a child did not come from the dishes that she liked, but rather the ones that she hated. In the same way, the left-handed amino acids might represent the leftovers that remained after the Martians had eaten their preferred right-handed form. Fortunately, she had packed both items among the Gulliver supply kit that she had smuggled from Earth and they were stored on the Rover. Minutes later, she inserted the right-handed form into the chamber, building a small pile next to the left-handed one.

Again, no crawling creatures came, but she watched in amazement as a small projection of yellow-red mineral crystallized out from the walls of the chamber, forming a finger that slowly extended toward the right-handed mound. It avoided the pile of left-handed material and, on arriving at the other one, moistened it and proceeded to absorb it. Janice focused her remote microscope on the edge of the mineral and turned her instrument up to highest magnification. Finally she saw it: the fragile towers, the tubes and the turrets, of the "city" of Mars. But this city was a busy one, for it resembled a factory, with a multitude of pores, partitions, pipes, and pumps. The Martians did not live in their microcity. They were the city.

Author's Explanation: The City in the Crater

In writing this story, I have tried to make the point that life elsewhere may be quite strange. In searching for it, we cannot rely on tests that would be valid for the microbes that we know on Earth. Instead we will want to keep our wits about us, sharpen our intuitions, and be ready to modify our plans

on short notice. Such circumstances almost demand an alert and respon-
sive human presence, not preprogrammed robots with a fixed set of tests
and an inflexible agenda. The mineral life form that I have invoked is pat-
terned after the one suggested by Graham Cairns-Smith for the origin of
life on Earth, but I have given it a taste for right-handed amino acids to
make it truly Martian rather than an organism that may have migrated to
or from Earth in an earlier era.

The bureaucratic ban on explicit life-detection experiments and the
reason provided for it were taken from an account by Gil Levin of his experiences in the 1990s, when he tried to have his instrument included on
scheduled Mars missions.

A number of books and magazine articles have advocated the human
exploration of Mars over the past two decades. President George Bush had
endorsed the idea, but support withered when the costs were estimated at
$450 billion to $500 billion. Recently Rubert Zubrin, an engineer by training, has described a plausible and less expensive ($20 to 30 billion overall)
basis for proceeding, in his book *The Case for Mars.*

In Zubrin's plan, the vehicle that will return the crew to Earth is sent to
the Martian surface well ahead of the crew. Another advance shipment de-
livers robots and the parts for an automated chemical plant that they will
assemble on the Martian surface. This plant then manufactures the fuel for
the return trip, using local ingredients that are available. After this task has
been carried out and verified, a four-person expedition sets out for Mars.
This group will remain on the surface for 500 days and use a pressurized
ground Rover for local expeditions that last up to a week. After a series of
such missions has been completed to prepare a site, a permanent base will
be constructed on Mars.

Space enthusiasts have embraced this scheme, but no broader movement has emerged. An effort has been made to establish one, however.
With Mr. Zubrin as the moving force, the Mars Society held its founding
convention in Boulder, Colorado, in August 1998. At least 750 people attended, and all of them got the chance to sign an 800-word declaration
that began with the words, "The time has come for humanity to journey to
Mars. We're ready." The current political situation concerning such an undertaking was summarized by the *New York Times* early in 1998: "Because
there is no political mandate for a human mission to Mars, NASA is approaching the possibility cautiously and with little fanfare. The agency is
sponsoring several low-cost research projects aimed at identifying technolo-

gies that could be used and readying blueprints for an endeavor, should one be called for."

The identity of the Martians will continue to be a mystery for some time to come, unless we get lucky on some low-budget approach. For all its color and controversy, though, the surface of Mars is not the most promising place to seek existing life. It lacks a key requirement for the Life Principle, a liquid or dense gas. As we have seen, the gas giants provide a dense gas atmosphere, but if we seek liquid, we will do best in the satellites that surround them. We must move farther out.

10

In the Realm of the Giants

And God said, Let there be a firmament in the midst
of the waters, and let it divide the waters from the waters.
And God made the firmament, and divided the waters which
were under the firmament from the waters which
were above the firmament; and it was so.
—*Genesis 1:6–7*

Ice World

One day, humans may stand on the surface of Europa, and their
view will be spectacular. With no atmosphere, the stars will be
fully visible against the blackness of space, as they were for astronauts on the Moon. The Sun, if it is up, will be much reduced in size—less
than a quarter of its width as seen from the Earth. But other objects will
compete for their attention. Jupiter, with its spots, bands, and whorls, will
hang huge in the sky, if our explorers have landed on the appropriate side
of Europa. On occasion, several other moons of Jupiter may appear,
swelling up at times to a size equal to that of our Moon in the sky on Earth.
But this scene will have to compete with the one on the ground.

Europa is an ice-covered world, but it will not be a skating rink. Much of it will appear as snow and ice do, after a snow plow has carried out its task. The icescape should be studded with low ridges, fissures, blocks, and crevasses, scattered about in all directions. To travel across this landscape will not be an easy matter. But if the landing team has chosen its site well, surface excursions will not be a priority. Attention will be focused on the realm below. If team members can penetrate through the ice, they may enter an ocean larger than those on Earth, one that has persisted for billions of years. If it truly exists, this sea world, these waters above our own firmament, offer the best chance to detect advanced life elsewhere in our solar system. In this case, the reality runs barely ahead of the dream, as the world was unknown until the seventeenth century, and its qualities were unappreciated until two decades ago.

Life in the Outer Reaches

Four of the wandering worlds known to the ancients, Mercury, Venus, the Moon, and Mars, now are members of the group called the inner solar system. Farther out lies the realm of the gas giants: Jupiter, Saturn, Uranus, and Neptune, and the multitude of worlds that circle them. Of the twenty-five worlds, including the Earth, of diameter 1,000 kilometers or more, sixteen are to be found in this region. Two members of this group that were unknown before the seventeenth century deserve our special attention: Europa, which circles Jupiter, and Titan, a moon of Saturn. Both of them claim the highest priority in our search for the Life Principle. We will start with Europa, which is closer and was among the first worlds added to the list known by early societies. We can begin by returning to our imaginary Museum of the Cosmos, and to the room that holds our model of Jupiter.

On Level 7 of this Museum, objects are modeled one ten-millionth of their actual size. The Earth would be a bit less than our own height, while Jupiter would be represented by a striped ball some 15 meters tall. To add the satellites first discovered by Galileo in 1610, we would need four beach-balls, varying in size from 31 to 53 centimeters. They would be placed from about half a city block to more than two city blocks away from the model of Jupiter, so we will require an entire floor of a large building to display this model of the Jovian system.

The Florentine astronomer Galileo Galilei, whom we met earlier, discovered the four worlds in January 1610, using a homemade telescope.

Jupiter has a number of additional moons and a wispy ring, but they were too small to be visible with his telescope. In our display, the largest of the remaining moons will be the size of a marble.

Galileo published his observations that March, commenting that

> I should disclose and publish to the world the occasion of discovering and observing four Planets, never seen from the beginning of the world up to our own times. . . .
>   I therefore concluded, and decided unhesitatingly, that there are three stars in the heavens moving about Jupiter, as Venus and Mercury around the Sun; which was at length established as clear as daylight by numerous other subsequent observations. These observations also established that there are not only three, but four, erratic sidereal bodies performing their revolutions around Jupiter.

Galileo proposed to name them the Medician satellites to honor the grand duke of Tuscany, Cosimo II de Medici, and he suggested names after members of that family. This idea did not catch on, and the four are now collectively called the Galilean satellites, in honor of their discoverer. But the worlds had also been discovered independently at about the same time by the German astronomer Simon Marius. Marius, at the suggestion of his celebrated colleague Johannes Kepler, suggested the individual names that they now bear: Io, Europa, Ganymede, and Callisto. He described the reasons for his choices as follows:

> Jupiter is much blamed by the poets on account of his irregular loves. Three maidens are especially mentioned as having been clandestinely courted by Jupiter with success. Io, daughter of the River, Inachus, Callisto of Lycaon, Europa of Agenor. Then there was Ganymede, the handsome son of King Tros, whom Jupiter, having taken the form of an eagle, transported to heaven on his back, as poets fabulously tell. . . . I think, therefore, that I shall not have done amiss if the First is called by me Io, the Second Europa, the Third, on account of its majesty of light, Ganymede, the Fourth Callisto.

Not all of Galileo's contemporaries were thrilled by his heavenly discoveries, which included the mountains and valleys of the Moon. Florentine astronomer Francesco Silezzi reasoned that the number of planets (seven, including the sun) was equal to the number of orifices in the human head and the days of the week, and therefore innate in the universal scheme of things. He further commented, "Moreover the satellites are invisible to the naked eye and therefore can have no influence on the earth and therefore would be useless and therefore do not exist."

Above all, Galileo's advocacy of the Copernican system, in which the planets revolve around the Sun, gradually brought him into conflict with the Church. Until that point, his studies on such topics as falling bodies and pendular motion had earned prestige and honors for him, including professorships at Padua and Pisa. In 1616, however, he was warned not to support Copernican ideas, although he could still discuss them as mathematical hypotheses. Galileo then went into a period of retirement that lasted seven years. After the death of his antagonist, Cardinal Bellarmine, he resumed his activity, eventually publishing in 1632 his master work *Dialogue Concerning the Two Chief World Systems—Ptolemaic and Copernican*. His reward was a summons to Rome, to stand trial before the Inquisition on suspicion of heresy. As another scholar, Giordano Bruno, had been burned at the stake by that body in 1600, in part for his advocacy of Copernican ideas, this invitation was not a light matter.

To avoid a more severe fate Galileo was forced to declare that he "abjured, cursed and detested" the teachings in question. He was sentenced to imprisonment, but the Pope commuted this to house arrest, which endured until his death in 1642. He was not fully pardoned until 1992.

By that late date, another Galileo was already en route to continue the studies of the moons of Jupiter. This successor was not made of flesh, but rather of metal and plastic. A spacecraft had been named in honor of the Florentine and dispatched by NASA to the Jovian system. This ship was also continuing the work of more recent predecessors. Two *Pioneer* spacecraft in 1973 and 1974 and two *Voyagers* in 1979 had flown through the system and brought us our first detailed knowledge of the four worlds. Like the Greek dieties for which they had been named, the Galilean moons were diverse and stunning. An extended reconnaissance from Jupiter orbit was needed, and eventually carried out, but only after a set of ordeals that recalls those of the original Galileo.

Trials of the Second Galileo

The Galileo mission was first approved by Congress in 1977 and fully authorized in 1979. A January 1982 launch date was set, but technical difficulties delayed this until May 1986. In January 1986, however, the space shuttle *Challenger* blew up, and scheduled launches were put on hold, including the one for *Galileo*. The powerful *Centaur* rocket that was to be housed in the space shuttle and used later in the launch was judged too dangerous and was banned. The spacecraft could no longer be sent on a direct path to Jupiter. NASA scientists rescued the mission by devising an indirect route that compensated for the loss of power by making use of gravity boosts from the Earth (twice) and Venus. The distance to be traveled was multiplied by four, however, and the one-way trip would require more than six years. Finally, on October 18, 1989, the *Galileo* spacecraft was launched, but its troubles were by no means over.

The main umbrella-like transmitting antenna had been kept in a folded position during the Venus fly-by to protect it from solar radiation. Unfortunately, it could not be opened after that, most probably because its ribs were damaged. According to project scientist Torrence V. Johnson, "The best engineering judgement is that the ribs are permanently jammed, probably because of the loss of lubricant during the long truck rides the spacecraft took from the Pacific coast to the Atlantic in 1986, back to the Pacific when launch was delayed and again to the Atlantic in 1989." A secondary antenna was available, but this could transmit at less than 1 percent of the rate, and a portion of the data available from *Galileo* would never be retrieved. Problems also developed with the tape recorder used to store the data prior to transmission. Despite the setbacks, however, a happy ending emerged. The mission, originally scheduled for a 1997 shutdown, was extended to December 1999, and a magnificent, if tantalizingly incomplete, portrait of the four worlds has been assembled. One of them, Europa, is engrossing in terms of our search, but we can pause for a glimpse of the others.

Galileo's Family

We will start our journey at the edge of the Jovian system and travel inward, as if we were arriving by spaceship. The outermost of the large moons, Callisto, is perhaps the least interesting. Its density tells us that it is a bit more than half ice, with the remainder rock and metal. Its heavily cratered surface appears to be a mixture of ice and dark soil or rock. In the words of *Nature* magazine, "A black cliff shadow stretches across a ruined landscape of icy hillocks and small craters, partially buried by dark, smooth material."

Different opinions have been voiced concerning the extent to which the coldness of Callisto's surface extends within the planet. According to some scientists, the interior was never melted to any extent and remains as a mixture of ice and rock. Many planets are heated enough to melt their interiors, and heavy metals such as iron sink to form a core. Lighter rocks float above that core, and ice, if present, moves to the surface. This process, called differentiation, never took place within Callisto, or did so only partially. If this viewpoint is correct, then Callisto remains airless, frozen, and, for our purposes, uninteresting.

Other scientists have been more influenced by Callisto's magnetic field than by its frozen surface. On the basis of new data returned by the *Galileo* spacecraft, Margeret Kivelson of the University of California, Los Angeles, commented at a 1998 meeting that "we think a subsurface ocean is likely." Rather than linger over this unresolved issue, we will move inward to Ganymede, the largest of the Galilean satellites.

Ordinarily, at Ganymede, we would expect to find Callisto all over again. Both have about the same composition and are equally distant from the warmth of the Sun. Ganymede presents a cratered, airless, icy surface, deformed by ridges, faults, and grooves. But, as reported by *Science* magazine, "The moon's calm and icy surface conceals a hot and turbulent interior." A substantial magnetic field surrounds the planet, with a center of molten metal as the most likely cause. Ganymede has probably differentiated; one model of its interior, from the center out, showed 800 kilometers of molten metal, followed by a rock layer of comparable thickness, and then an ice layer that extended to the surface.

Why should Ganymede and Callisto be different? A key reason may lie in their relative distance from Jupiter; Callisto is 1.8 times as far away as Ganymede. Jupiter, of course, never developed into a sun, but it affects nearby worlds in another way. It tugs at them with the gravitational attraction of its huge bulk, stressing and flexing their solid materials and shifting liquids about. These interactions are called tidal forces, because the effect of the Moon in creating tides on Earth provides a familiar example. For tidal forces to act, however, the orbit of the moon around Jupiter must be eccentric and not perfectly circular. Possibly, an isolated moon could adapt in this way, but the movement of each of Jupiter's moons is further perturbed by the coming and going of the others, creating a small orbital eccentricity in each.

Tidal forces fall off sharply with distance, and the current position and eccentricity of Ganymede's orbit should not generate enough heat to melt its interior. Planetary scientists speculate, however, that its orbit may have been more eccentric billions of years ago, and the meltdown that took place at that time has not yet cooled off.

Whatever the cause of Ganymede's molten center may have been, its effects are real. If the interior is hot enough, the closer portion of the ice layer may have melted, affording an internal ocean. One crucial condition needed for the generation of life, that of a fluid medium, would be satisfied and the liquid would be the one familiar to us on Earth, water.

My colleague Gerald Feinberg and I suggested in a 1979 conference and in our subsequent book, *Life Beyond Earth,* that Ganymede would be a plausible place to search for life. Of course, no light would reach an ocean buried under 1,000 kilometers of ice. But Gary (as his friends called him) was aware that life had been discovered at deep sea hot vents on the ocean floor on Earth. These underwater hot springs can be found at places where the Earth's crust is thin and molten rock comes close to the surface (for example, near the Galapagos Islands in the Pacific). These vents have been explored using submersibles such as the Alvin, a small battery-powered research vessel that can carry a crew of three down to great depths.

Light does not penetrate to the deep sea bottom, but bacteria survive there by harvesting the energy present in the chemicals released in the vents. More advanced organisms such as tube worms, mussels, and clams live off the bacteria. Gary speculated that life could survive at the bottom of a Ganymede ocean as well, provided the necessary vents and chemicals were available there.

To validate such a speculation, of course, we would probably have to tunnel through the huge thickness of ice that covers the planet. Such thoughts have been set aside as a similar and much more accessible target exists in the Ganymede's neighbor, Europa. That moon has less water than its neighbors farther out, but it would be more subject to tidal heating by Jupiter. By contrast, the innermost planet of the quartet, Io, which we discussed earlier, suffers from tidal overheating. It has little water, and active volcanoes continually spurt sulfur compounds and molten rock onto its mottled surface.

An Ocean Even Farther Away

The risk one runs in exploring a coast, in these unknown and icy seas, is so very great, that I can be bold though to say that no man will ever venture further than I have done. . . . Thick fogs, snow

storms, intense cold and every other thing that can render navigation
dangerous must be encountered; and these difficulties are greatly
heightened by the inexpressibly horrid aspect of the country; a
country doomed by Nature never once to feel the warmth of
the sun's rays, but to lie buried in everlasting snow and ice.
—*Captain James Cook, 1775, commenting on his explorations in Antarctic waters*

The photos released by the *Voyager* fly-bys of 1979 showed Europa to be a
very bright world, one that was relatively featureless and almost uncratered,
compared to Ganymede and Callisto. Some areas were slightly less bright,
however. When these were given artificial colors in the photos, the surface
appeared as an intricate network of straight and modestly curved lines, one
reminiscent of reinforced glass that has been shattered or some areas of the
later paintings of the mid-twentieth-century American abstract artist, Jack-
son Pollack. As Europa had shared the same history of battering by meteorites as its neighbors, its relative absence of craters provided a puzzle.

One solution was that the craters had been formed but then swallowed. Some planetary scientists speculated that an underground ocean lay
just a few kilometers below the icy surface, providing it with a plastic quality. That ocean could be as much as 100 to 200 kilometers deep, with a volume greater than all of the combined oceans of Earth. The icecap above
it, exposed to the vacuum of space, was a frigid −162°C (−292°F).

But the existence of this ocean was by no means certain. Other scientists suggested that a layer of soft ice might produce the same effect as the
supposed ocean. Or perhaps an ocean had formed early in Europa's history and still existed at the time when most of the cratering occurred. But
that ocean might have frozen solid billions of years ago. If the rate of
crater formation since then had been very slow, Europa's face might have
an artificially young appearance.

Little progress was made in this debate until the *Galileo* spacecraft
began a series of repeated fly-bys at close range, starting in December
1996. With each new passage, more data have been produced that support
the idea of an ocean beneath the ice, and the skeptics have gradually come
over. The media and some scientific journals have taken up the theme with
headlines such as "An Ocean Emerges on Europa" and "New Images Hint
at Wild and Wet History for Europa."

For example, a press release of March 1998 heralded the closest Europa fly-by yet, with details only a few meters across visible in photos.
Three separate lines of evidence suggested an ocean, perhaps a salty one at

least 100 kilometers deep. The crater Pwyll was cited for its strangely shallow appearance. The central peaks of the crater were over 600 meters tall, higher than the rim, but the basin looked as if it had been filled in by water, which then froze. The debris surrounding the crater was readily visible, which suggested that Pwyll was relatively new.

A series of parallel ridges in another area appeared to have been formed as sheets pulled apart repeatedly and liquid rushed in to fill in the cracks. Yet another region featured rugged terrain that may have been created when a field of icebergs, frozen in place, was rotated and fractured. Comparable structures have been formed in regions of the Arctic where sheets and blocks of ice cover the open sea.

These observations, and many others, have painted a convincing but not yet conclusive case for the Europa ocean. One key piece of evidence was still lacking that would provide the "smoking gun" (or perhaps "cold knife" would be the better metaphor): the observation of a water eruption. It may yet appear during the last *Galileo* measurements. If not, we must await another visit. If we can assume for now that the ocean does exist, then we can move on to the more vital question: What will we find in it?

Arthur Clarke's Tale

To the surprise of the entire world, a crew of five Chinese astronauts landed their spaceship safely on Europa and claimed title to the planet for their country. Their chosen landing place was a rock outcropping immediately adjacent to the Grand Canal, a 1,500 kilometer long crack in the surface of Europa's ice. The Grand Canal had formed relatively recently; it was filled with a crust of newer ice that was only a few fingers thick. The crew of the *Tsien* had punctured it and was pumping water into their propellant tanks when disaster struck. The last survivor, Professor Chang, told the tale by radio before he too perished.

The ship had been illuminated by a bright string of colored lights so that it resembled a Christmas tree. Suddenly something emerged from the icy crack and climbed onto the surface, moving quite slowly toward the *Tsien*. It looked "like huge strands of wet seaweed, crawling on the ground. . . . It was freezing solid as it moved forward—bits were breaking off like glass—but it was still advancing toward the ship, a black tidal wave, slowing down all the time."

Chang continued, reporting how the creature had crawled onto the spaceship, carrying tons of ice with it. Apparently, it had been attracted by the lights. As a result, the *Tsien* toppled over and was wrecked. The Professor then provided a closer description:

> Imagine an oak tree—better still a banyan with its multiple trunks and roots—flattened out by gravity and trying to creep along the ground. . . . Then I saw that large buds were forming on the branches. Delicate, beautiful colored membranes started to unfold. . . . There were scores of big flowers, in various states of unfolding. . . . But they were freezing-drying as quickly as they formed. Then, one after another, they dropped off from the parent buds. For a few minutes they flopped around like fish stranded on dry land—and at last I realized what they were. Those membranes weren't petals—they were *fins*, or their equivalent. This was the free-swimming, larval stage of the creature. Probably it spends much of its life rooted on the seabed, then sends these mobile offspring in search of new territory.

Chang felt that the creature was likely to die if it remained on the surface, so finally he disconnected the lights and kicked at it until it retreated into the crack in the Canal. Had he taken this step earlier, he could have prevented the tragedy.

This account was taken from Arthur Clarke's 1982 novel, *2010: Odyssey Two*, which was subsequently made into a film. At the end of the novel, a representative of an advanced civilization informs the human race that it may inhabit all the Jovian moons except this one: "All these worlds are yours save Europa." Life on that world must be free to develop without human interference. Clarke acknowledged a 1980 article by journalist Richard Hoagland for the suggestion of life in a Europan ocean.

Less than two decades later, in 1996, I sat in a spacious Mexican restaurant in San Juan Capistrano, California, attending a banquet dinner for the more than 100 scientists who had attended the first Europa Ocean Conference. A highlight of that evening was a live satellite telephone call to Clarke at his home in Sri Lanka. He addressed the audience, inquired about the location, weather, and agenda, and then answered questions. Above all, he wanted to assure us that the creature he had described in *2010* intended no harm to humans but has simply been drawn by the lights.

That conference had been called by an astronomical institute to review the evidence concerning the possibility of liquid water on Europa, immediately before new data from the Galileo mission began to pour in. Many planetary scientists were present, such as Torrence Johnson and Professor Steven Squyres of Cornell, an expert on Europa. But a very different group of scientists had been invited as well. A substantial number of marine biologists attended, including John Baross and John Delaney, of the School of Oceanography of the University of Washington. During that same dinner, Delaney showed us an impressive videotape featuring close-ups of the smokers, chimneys, and other deep sea hydrothermal vent features where marine life congregates. The sense of the conference was clearly that a Europan ocean existed and that life was likely to be found at the bottom of it.

The Life Principle: A Crucial Test

The preceding two assumptions can be coupled together readily, but they differ enormously in their implications. Geological and astronomical circumstances have determined whether or not an ocean exists on Europa. If one is there, then its presence will explain many features of the surface to the satisfaction of planetary scientists, but no greater meaning can be drawn from its existence. It would, however, provide an important test for the Life Principle, with implications that extend throughout the universe.

That principle specifies that a liquid (or dense gas), an energy supply, a suitable set of chemicals, and an interactive energy source are needed to set self-organization on its course. If the Europa ocean does exist, as seems likely, then it probably has been there for billions of years. The availability of chemicals and energy would then be the critical factors. We may, of course, encounter an ocean of pure water, ready to be bottled and sold in health stores on Earth. But some evidence has already been collected that indicates the presence of mineral salts. Traces of simple organic compounds have been detected on the surface of Ganymede and Callisto, so it is reasonable to presume that they have crept into Europa's ocean as well. The question of a supply of useful energy is more troublesome. Tidal forces may have melted the ice, but the heat gradients that they provided may not be usable for self-organization. Chemical energy, as provided by deep sea vents or other sources, may be much more appropriate. We can only hope that they were available and get on with our search.

We truly do not know what we will find. If the pessimists that I have called the Sour Lemon School are correct, then a vast supply of mineral water will be our only reward. If the Life Principle has been at work, then we may be as surprised as Professor Chang, but hopefully with less lethal results. As we have learned on Earth, 4 billion years of evolution can produce striking outcomes. It may be that all we need to do is land on the surface, and Europan life will crawl out in search of us. If that life has been abundant, we may not even require a lander to detect it. Physicist Freeman Dyson has suggested an easier and less expensive alternative for seeking Europan life:

> Every time a major impact occurs on Europa, a vast quantity of water is splashed from the ocean into the space around Jupiter. Creatures living in the water far enough from the impact have a chance of being splashed intact into space and quickly freeze-dried. Therefore, an easy way to look for evidence of life in Europa's ocean is to look for freeze-dried fish in the ring of space debris orbiting Jupiter. Sending a spacecraft to visit and survey Jupiter's ring would be far less expensive than sending a submarine to visit and survey Europa's ocean. Even if we did not find freeze-dried fish in Jupiter's ring, we might find other surprises—freeze-dried seaweed or a freeze-dried sea monster.

As the present climate in Washington, D.C., for planetary science does not favor such creative imagination, I suspect that the future exploration of Europa will proceed in a much more conventional way. Scientists at NASA have started to plan for an orbiter that will circle Europa and, by using radar and other techniques, confirm (or deny) the presence of an ocean. If one should be present, then its depth and the thickness of the ice that covers it will be measured. A group at Jet Propulsion Laboratory is preparing for a follow-up mission that will place a lander on the surface. The group is developing a "cryobot," which will melt its way through the ice layer, spooling out a communication cable as it goes down. When the ocean is reached, a tethered "hydrobot" submersible will be deployed to make measurements and take a look. Fortunately, this combination of devices can be tested on Earth before it is sent so far from home.

Deep under the ice of Antarctica lies a huge trapped body of water called Lake Vostok. It has the dimensions of Lake Ontario but is covered by 3.7 kilometers of ice. Russian scientists have drilled down close to the lake but stopped, for fear of contaminating it. This site may provide an ideal testing ground for the technology intended for Europa, before we send it off.

Robot explorations may well settle the question of Europan life. Even if we do not send spacecraft for a cruise out in Jupiter's ring, the first landers may find chemical or fossil evidence of life that has splashed onto the surface. Or the first photographs of the ocean may resemble the scene that confronts us when we visit the aquarium. If our Martian experience thus far serves as any guide, however, the first clues will be ambiguous. Should a deeper search be required, then we will want to check out the scene in person. If the robots succeed, then our scientific curiosity may take us out there anyway, and if life is discovered then we will certainly want to examine its habitat in person. Unless, of course, Arthur Clarke's advanced civilization forbids it.

Once we have traveled out as far as Jupiter, we will not want to stop. The next system out offers an even more glorious spectacle and a very different way to search for the Life Principle.

Petroleum World

When a *Tourist Guide to the Solar System* is eventually published, the view of Saturn from space, or from the surface of an inner satellite, will undoubtedly receive the maximum number of stars. The rings of this gas giant planet have become a symbol for space exploration. Yet they were unknown before the seventeenth century and only fully appreciated with the spectacular *Voyager I* fly-by in 1980. We have learned that they extend in multiple complex bands over a width of some 275,000 kilometers but are probably less than 100 meters thick. Sir Fred Hoyle may perhaps believe that they are composed of frozen bacteria, but most scientists think that they are made of rocks. Most of the ring components range from the size of marbles to basketballs, but some tiny moons are present as well. Despite their intricacy and beauty, our goal here is not sightseeing but a search for the Life Principle, so we will turn our attention to Saturn's largest moon.

Titan ranks second only to Ganymede in size among the satellites of our solar system, and it is larger than the planet Mercury. It differs from all of the other moons, however, in one important way: It has an appreciable atmosphere. If we exclude the gas giants, then only Venus and Earth keep it company in this category. Orbiters have now given us a map of Venus, but Titan has kept itself heavily veiled, teasing us as if it were an Oriental dancer. The *Voyager* fly-by showed only the orange haze of the atmosphere, without a single glimpse through the clouds. We did not see the face of the planet, but we learned something of its temperament.

Before that time, writers could imagine what they wished, and Kurt Vonnegut did just that in his 1959 satiric novel, *The Sirens of Titan:*

> The atmosphere of Titan is like the atmosphere outside the back door of an earthling bakery on a spring morning.
>
> Titan has a natural chemical furnace at its core that maintains a uniform air temperature of sixty-seven degrees Fahrenheit.
>
> There are three seas on Titan, each the size of earthling Lake Michigan. The waters of all three are fresh and emerald clear. . . .
>
> There are woods and meadows and mountains. . . .
>
> Titan affords an incomparable view of the most appalingly beautiful things in the Solar System, the rings of Saturn.

The selection that I have chosen is remarkable for its contrast with Titan's reality, as revealed by *Voyager* and observations from Earth. The atmosphere is there all right, with the surface pressure one and one half times that of Earth. But we would not care to breathe it. An astronaut on the surface who opened the helmet of his suit to sniff the bakery smells would immediately be frozen and asphyxiated, as the temperature is a subfrigid $-180°C$ ($-292°F$), and the air lacks oxygen. If any vapors warmed up enough for him to sense them before he expired, then he might register the odor that permeates a oil refinery. Nitrogen is the major component, as it is on Earth, but the similarity ends there. Hydrogen gas, once used here to fill balloons, is also part of the air of Titan, and most of the remainder is made of hydrocarbons, substances made of only carbon and hydrogen, that we encounter in natural gas and petroleum.

There may be seas on Titan, though we have no idea of what their number or size would be. But they could not be made of water, as it is much too cold there. In that topsy-turvy world, the mountains would be made of ice, while the seas would contain hydrocarbons such as methane and ethane. These chemicals may be gases on Earth, but they fill up the lakes of Titan. They could hardly be called "fresh" though, unless you are used to applying that word to the contents of your fuel tank, and "emerald clear" would not be expected under an orange sky. The tarry materials that remain behind after one has distilled off the lighter fractions of crude oil may also discolor the icy surfaces and darken the sea bottom on that world.

Scientists still debate whether bodies of liquid do exist on the surface of Titan. In the 1980s Carl Sagan, Jonathan Lunine, and others proposed that an ocean may cover much of that world. But if that were the case, then tidal forces would have acted to make Titan's orbit perfectly circular. Nature has given us a different message: The orbit is irregular. Radar observations have suggested that some of the surface is exposed ice, and the Hubble space telescope has returned data revealing lighter and darker areas on Titan. If the darkest areas were separate seas, then at best 10 percent of the planet could be covered by them.

Returning to the Vonnegut quote, we can see why woods and meadows would not be expected, though mountains of ice may exist. Even the last sentence of the quote is mistaken. The beauty of Saturn's rings could not be appreciated on the surface of the planet because of the thickness of the haze, which prevents any observation of Saturn.

This circumstance was also ignored, for poetic reasons, by the artist who prepared a NASA poster, "Amazing Saturn." His view featured an imagined Titan landscape in the foreground, with Saturn in the sky above. In other ways that poster represented as good a guess as any as to what the landscape may be like. A crater lake sits between rugged, barren cliffs, with the entire scene—cliffs, lake, and sky—rendered in muted colors of orange, dark brown, and white.

The artist has included one additional feature that will be visible for a period of only two hours, in the year 2004: An apparatus is descending by parachute from the sky. That event will take place when the *Huygens* probe is released near Titan by the *Cassini* Saturn orbiter. The spacecraft and its probe are named for the Dutch physicist Christianus Huygens, who in 1645 discovered Titan, and the French-Italian astronomer Jean Dominique Cassini, who first observed four other moons of Saturn and a gap in its system of rings.

The unique properties of this veiled world have attracted the interest of planetary scientists and helped secure funding for *Cassini*. An article in *Chemical and Engineering News* called it the "last of the big planetary missions" and added the following explanation: "The Cassini mission was conceived in the early 1980's, when elaborate all-or-nothing projects (which included the Galileo mission to Jupiter) were the status quo. In the 'faster, better, cheaper' climate of today's NASA, a project as large as Cassini would never be funded, some scientists say." The overall cost, some shared by foreign collaborators, was estimated at $3.3 billion, hardly a thrift conscious item.

The term *faster* would also not apply to a trip that required seven years for arrival at its target, after an October 1997 launch. Linda Spilker, a NASA deputy scientist for the mission, explained to me that a direct routing would have had the spacecraft arrive with too much speed to achieve orbit. Instead, a *Galileo*-like trek with two gravity assists by Venus and one by Earth was necessary. That last fly-by had triggered prelaunch protests from opponents of the mission, who feared that an accident involving the craft's plutonium power source might contaminate Earth's atmosphere. Its supporters, of course, included not only planetologists, but also those of us who are motivated to learn more about our origins and the place of life in the cosmos. Titan offers a unique opportunity to harvest information about self-organization by chemicals.

If a Life Principle exists, then surely it has had a chance to weave its magic on Titan. Hydrocarbon lakes may decorate the surface, but if liquid should be lacking there, then it may be lurking just below the porous crust. Carbon compounds are abundant on Titan, and we know from our own example that carbon chemistry can serve as a basis for self-organization. Many of the energy sources cited for the early Earth are present there as well, though in reduced amounts. Sunlight filters down through the atmosphere, and some reaches the surface. In the upper reaches of the atmosphere, solar radiation promotes chemical reactions, and the products can descend to the surface, providing possible sources of energy. The entire set of calamities that is cited in connection with the start of life on Earth—thunderstorms, volcanic eruptions, and meteorite impacts—undoubtedly have disturbed Titan as well. If energy can be thought to be the true "food" for life, then we can claim ignorance of the exact dishes that Titan prepares and the size of the portions, but we can be fairly secure that there are several different items on the menu.

In my eulogy on the virtues of Titan, I have neglected to remind you of the dreadful cold that grips the planet. Surface temperatures there are less than 100°C (or 180°F) above the coldest possible temperature, which is called absolute zero. Scientists who feel that life can only function with a set of chemicals very similar to our own will lose interest at this point. The reactions that I teach to biochemistry students would take place at an insignificant rate in the frozen climate of Titan. In fact, most of our bio-chemicals would refuse even to dissolve in the chilled petroleum sea.

But suppose we ask different questions. Do some chemicals exist that will enter happily into that sea? Are there some processes involving much weaker bonds that could still take place at a reasonable rate in that environment? Can these reactions interconnect with one another in a way that sets up catalytic cycles? Can the system absorb energy in a way that moves it in the direction of greater organization? If the answer to these questions is yes, then complexification can begin its work. In the words of planetary scientist Toby Owen and his colleagues, "Titan offers us an enormous natural laboratory in which a variety of experiments in chemical evolution are taking place today."

The pace of the process may have been quite slow in the cold, but it has had a lot of time in which to develop. We may observe the Life Principle at work in its earliest stages, and it is exactly those stages that we understand the least. Yet our ignorance of what we can expect may make it harder for us to appreciate what is there.

Titan provides a more difficult challenge than Europa. It may be easy for us to locate Europan life. We may see moving shapes in the internal ocean that are obviously alive, or find their dead remains ejected onto the surface. We should not expect seaweed creatures or even tube worms to crawl out of the murky liquid pools on the surface of Titan. A much closer and more detailed analysis will be needed. We must seek complex mixtures of chemicals in which some items are present in unexpected excess while their close cousins are absent. A chaotic mixture that contains almost everything, as we found in the Murchison meteorite, will signal a negative result.

The more promising mixtures may be localized and perhaps segregated within compartments. They may be located only in particular areas on Titan (for example, near an energy source). Only after a detailed survey will we be able to learn whether some process that leads in a biological direction is taking place.

Traditionalists feel that nothing of that sort should be expected. In the words of origin of life chemist Jim Ferris and his colleague David Clarke, "It is generally accepted that Titan can be used to simulate planetary scale chemical activity and surface-atmosphere interactions in the absence of life's influence. Any further exobiological use meets with resistance as the universally accepted ingredient for life has not been detected on Titan: liquid water."

Presumably, those of us who disagree that this ingredient is essential have been banished to another universe. But there may be a prize on Titan for the water-or-nothing group as well. Its surface will be sculpted from solid ice and snow, just as it is made of rock and soil on Earth. The hot interior of the Earth contains molten rock, however, and on occasion it spurts onto the surface and then solidifies. We do not know whether Titan contains substantial internal heat. If it does, then carbon chemistry in water may have been bubbling along merrily for some time in the underground. We can sample it by inspecting the eruptions, which would have been preserved quite nicely by the extreme cold.

We could hope for present or past water volcanoes on Titan, but no sign has emerged yet to encourage us. But another option for aqueous chemistry exists there, which is more likely. Carl Sagan has estimated that any spot on Titan has had a 50-50 chance of being hit by a meteorite and melted at one time in its history, given the record of cratering that we can observe on other moons of Saturn. A temporary pond or lake of water would be formed that could last 1,000 years before freezing over. In a lecture at my university, Jonathan Lunine extended that time to 10,000 years for a crater that was as large as 10 kilometers across.

The circumstances in that temporary lake would have much in common with those in the famous origin of life experiment performed by Stanley Miller and Harold Urey in 1953 (see Chapter 5). In that experiment, some amino acids were produced when an electric discharge was passed through a mixture of methane, water, hydrogen, and ammonia. The first three substances are present on Titan, and ammonia is believed to be a component of the ice there. Over the past billions of years, we would have the opportunity for a large number of Miller-Urey experiments to be initiated, run for a variable length of time, and then preserved by freezing.

As I indicated earlier, the preparation of simple amino acids does not represent a great leap forward in self-organization. The key to the riddle of life's origin lies in the subsequent steps. If some abrupt and unlikely stroke of luck was required on Earth to produce a molecule that could copy itself, we would be unlikely to hit the jackpot again in the limited opportunities on Titan. If interacting chemical cycles were involved, as suggested by complexity theory, then we might be fortunate enough to see such a system preserved in its initial stages.

This issue, and the question of chemical evolution in existing hydrocarbon lakes, will not be settled by the *Huygens* probe. That apparatus will determine the simpler ingredients of the atmosphere more accurately and perhaps provide us with some spectacular photographs as it parachutes down. It is equipped to function on the surface for up to half an hour after it lands, but it cannot guide itself to an interesting site. According to Jonathan Lunine, it may touch down on the edge of an area that is relatively bright, as seen in Hubble telescope images. This region may represent a stretch of comparatively clean ice, at an altitude higher than the hydrocarbon-stained lowlands. The *Cassini* spacecraft will enjoy a longer life, which will involve multiple fly-bys of Titan. Two of its tasks will be to determine the nature of the surface and to prepare a map. It is hoped that these initial results will whet our appetite for further explorations.

The insights that we gain from Titan about the origin of life will not come quickly. A close inspection of local environments may be needed. If we carry this out by robot, then we may need to digest the results of each mission before we plan the next one. If each sequence requires some time for planning and funding, followed by a wait for the next favorable launch opportunity and five years while the spacecraft is en route, then the extended process could eat up much of the next century.

I would rather dream of the time when the spirit of human adventure has brought astronauts out as far as the Saturn system. Such a mission could provide a highlight of twenty-first-century exploration. Some scientists could study the magnificent rings from the close-up platform of an airless inner moon. At the same time, geologists and chemists would carry out a treasure hunt on Titan's surface. They would sample the lakes of oil and probe the unknown frozen lands, under the hazy orange-red sky. The treasure that they sought would not be of gold or diamonds, but of insight into our origins and the forces that may produce life elsewhere in this universe.

11

Signs of
Ancient Visitors

On the afternoon of July 4, 1997, a standing-room-only crowd filled the 1,800-seat Main Hall of the Pasadena Center in California, to watch television coverage of the Mars *Pathfinder* lander on a 25-foot video screen. Scientists came in person from the nearby Jet Propulsion Laboratory to add their live presence to the coverage. This event was part of a much larger Planetfest 97 celebration that filled the center from July 3 to July 6 and attracted 20,000 visitors overall. Those who attended were entertained and informed by lectures, panels, and exhibits from past and future space missions. They could also get a glimpse of the celebrated Martian meteorite, ALH84001. As described in the *Planetary Report*, "The three-day celebration marked a new highpoint in our fantastic journey beyond planet Earth." Interest in *Pathfinder* extended far beyond Pasadena, of course. *Time* magazine, for example, devoted its July 14 front cover to a photograph of the Rover *Sojourner* against a backdrop of the Martian landscape.

At the exact same time as the Planetfest 97 gathering, an alternative celebration was being held in Roswell, New Mexico, with an attendance of 40,000, twice that of Planetfest. This gathering, called Roswell UFO Encounter '97, was convened to mark the fiftieth anniversary of the alleged crash of a flying saucer there in the summer of 1947. Parades, exhibits, speakers, and dances all honored that earlier event. The celebration at Roswell, just like the one at Pasadena, attracted the interest of a substantial fraction of the American public. The *Time* magazine front cover for June 23, 1997 (three weeks before the *Pathfinder* landing), featured the imagined face of a bald humanoid alien, with a large forehead and a childlike body (which was shown within the magazine). The cover article reported a poll of 1,024 adult Americans, in which 22 percent believed that "intelligent beings from another planet have been in contact with human beings" and 17 percent agreed that "intelligent beings from other planets have abducted human beings to observe or experiment on them."

Roswell has become a shrine for this belief system because of its famous incident. Advocates claim that an extraterrestrial spaceship actually did crash there in 1947, and that its crew was killed. Agents of our government confiscated the wreckage and the bodies of the aliens and covered the incident with a made-up story about a crashed weather balloon. A number of books have described these events, and an old film of the autopsy of a supposed extraterrestrial has appeared on television. Science fiction films have started from this point and expanded the theme. In *Independence Day*, for example (in case you haven't seen it), Earth is attacked by extraterrestrial invaders who arrive in enormous spaceships. The Roswell saucer had served in one of their advance scouting forays. Our military has repaired it and kept it in an underground base for fifty years, without the knowledge of the various presidents. We now fly it ourselves as part of a counterattack that defeats the enemy.

Harvard professor John Mack and others have interviewed many individuals who claim to have been abducted in alien spaceships and then released. A *New York Times* account, for example, described Mack's meeting with a middle-aged technician. The technician claimed that "he was seduced by a 'female being' who had 'long silvery hair with large black eyes without pupils or irises' and who explained to him that she needed his sperm to 'create special babies.' "

I have traveled a small distance along a path that has been much trampled down by others to make a point. The belief systems celebrated by the festivals in Pasasdena and Roswell are incompatible. The Pathfinder mission represented a small step forward in our exploration of Mars, one important purpose of which is to determine whether life exists elsewhere in the universe. If the Roswell account is correct, that question has already been settled with a resounding yes.

Even if we established that a Life Principle does exist, and that extraterrestrial life is abundant, it wouldn't necessarily follow that other *intelligent* beings are nearby. Life started relatively quickly on Earth, but a much greater period, perhaps one-third of the time since the Big Bang, was needed for it to evolve to the point where it could cross interplanetary space. Yet that accomplishment may be so difficult that we are among the first to achieve it. We have no information about the extent to which intelligence exists in the universe. But Roswell advocates settle that problem as well, in the same breath with which they tell us that life has started elsewhere. Not only do other intelligent beings exist, but they are well ahead of us in technology and spend their time toying with us.

If their accounts were correct, then the exploration of Mars and Europa would have as much importance for us as a biological trip to Brazil to classify additional species of butterflies. Such research might have justified the expenditure of thousands of dollars, but not the billions spent in the search for life on Mars. The development of better rockets would also have been a waste of money, as they exist already and are being used by the aliens. Why should we spend enormous sums to fly to the Moon when we could charter an extraterrestrial taxi? If the Roswell incident, the Mack interviews, and the rest should be accurate, then our crushing overall priority would be to establish friendly contacts with the aliens and learn what their intentions are toward us.

The evidence, in fact, points strongly in the other direction. A detailed argument has been presented to establish that Roswell represented the crash of a military-financed balloon, one designed at my own university in a classified study called Project Mogul. A 230-page military report, which is available on the Internet, has been issued describing U.S. Air Force involvement in the project. Of course, books and articles arguing the opposite point are abundant, but that does not mean that we should leave the situation in limbo, as we have had to do for the Viking Labeled Release experiment, for example.

We have made good use of the rule, "extraordinary claims require extraordinary confirmation," and this dispute provides us with almost the textbook application. No quantity of personal accounts, photographs of objects in the sky, or sketches of humanoids can ever establish the claim that aliens are here, as too much hangs upon it. Other sorts of evidence, not yet forthcoming, would do quite nicely, however. For example,

1. The recovery of an artifact of a technology well beyond ours. Think of the impression that a time traveler to the Middle Ages could make upon the residents with a laptop computer.
2. The prediction of a totally unexpected astronomical or geological feature, taken from the aliens' maps.

3.   The return of an extraterrestrial biological specimen. A cutting taken from one of the potted plants that the aliens brought from home would serve well enough, though a hair or flake of skin from a humanoid would be even better.

The last category would be superb for the purposes of this book. I have wondered why the Roswell advocates claim that the aliens are extraterrestrials. This seemed plausible in the 1830s, when the Moon was thought by many to be inhabited, or at the start of this century, when Lowell's Mars was available. I have argued for the possibility of primitive life elsewhere in our solar system, but no alternative home for intelligent life appears to exist within it. By default, believers in flying saucers have had to presume that a much larger mother ship, capable of interstellar travel, is hidden somewhere out of sight.

Unless, of course, the aliens came from Earth rather than from abroad. It seems far more likely that a race of humanoids would have developed as an extra branch on our family tree on this planet and then sequestered themselves out of sight to avoid conflict. This would explain their appearance (somewhat like the science fiction visions of our own future evolution) and their interest in crossbreeding, as well as the scarcity of sightings of huge mother ships as compared to smaller saucers.

A tissue sample would reveal whether the "extraterrestrials" are DNA-based and as related to us as are, say, the Neanderthals, or whether they truly have an alien biochemistry and a separate origin. But I suspect that no such sample will ever be produced, neither smuggled out of a secret government alien mortuary nor pilfered during some abduction of a fertile human on an orbiting saucer. They will not be produced because, most likely, they do not exist.

Why, then, have I brought this narrative back to Earth from the far reaches of the solar system to consider such matters? I have done so because I want to argue that a search for alien artifacts on other worlds of our solar system is worth our while. To understand why, we must take a brief look at our own past and our likely future development.

Our Future as a Mirror of the Past

In Chapter 2, we compressed the history of the universe into a six-day calendar, to match it to the time scale used in the book of Genesis. In our scientific version of Creation Week, the Big Bang took place just after midnight on Monday, but the Earth did not come into existence until Thursday afternoon, and life did not appear here until early on Friday. Written human history fell into the last second of this week, and the Space Age has taken place within a small fraction of that second.

Both the history of life on Earth and of human culture on this planet have been marked by numerous setbacks and disasters (for example, the great Permian extinction of 250 million years ago and the decline of civilization in Western Europe during the Dark Ages). These collapses have been counterbalanced by spurts forward, such as the Cambrian explosion and the recent development of a global technological society.

The Apollo expeditions represented a remarkable expansion of the range of our biosphere. At that time, it seemed possible that this foothold would be enlarged immediately. As we noted in Chapter 1, Arthur C. Clarke had predicted "numerous inhabitants" and a childbirth on the Moon by the end of the twentieth century. Instead, the initial three-year advance was followed by a twenty-five-year reversal, which still continues. Conceivably, the Apollo landings could represent a highwater mark, like Pickett's charge at the Battle of Gettysburg, whose further advance would never be resumed.

A nuclear war, an asteroid collision, or some other catastrophe could reduce our global civilization here to a level from which it did not have the time to recover before longer-term effects made our planet uninhabitable. A future of this type might be palatable for the Sour Lemon School, but I think that we can do much better.

If we regain our energy and continue our expansion into space, then we will explore our solar system, at first with robots and then through human bases and colonies. Some observers have argued, of course, that we should settle for robot exploration alone. There is no need for humans to venture farther into space. Possibly, they may prevail, and our future will again follow the Sour Lemon line.

Even if we should colonize the solar system, we can expect the same arguments to come up again concerning further expansion. Return-trip travel within our solar system will be possible for adventurous humans, just as a trip to China was possible for Marco Polo and a sea voyage around the globe for some of Magellan's crew. To voyage to other star systems would require new technology and the willingness of humans to embark on expeditions in which only their descendants would arrive. We might achieve this, or we might once again have the option of settling for robot exploration to satisfy our curiosity, before we turn inward. A 1994 gathering of space scientists was convened to consider the robot exploration of nearby stars; it produced comments such as "there has been incremental advance in every aspect of this field" and "we are within striking distance of proposing a precursor mission." A similar meeting in 1998 remained optimistic.

Alternatively, humans might embark upon an expansion that results in our colonization of much of the galaxy. Although many technical problems must be overcome before we can construct starships that could attain some fraction of the speed of light, a variety of solutions have already been put on the table. Astronomer Ian Crawford has reviewed the prospects and reached the following conclusion:

> Neither the technology nor the economic base necessary to achieve rapid interstellar flight exists at present, and will not exist for decades or centuries to come. On the other hand, it is also clear that interstellar travel violates no physical law, and is therefore a legitimate technological goal for the distant future. We have already achieved the capability for very slow interstellar travel, and there is no reason to believe that the much higher velocities considered here will forever be out of reach.

Once the technology was in place, the human race could spread out. Physicist Sebastian van Hoerner has described such a future:

> Larger groups of thousands of volunteers may decide to take off in huge "mobile homes" on interstellar trips lasting many generations, finally colonizing the planets of other stars. And, after a certain settling time on each planet, the same cycle may repeat, leading again to mobile homes on another interstellar trip to the next stars and their planets. In this way we would have started a wave of stepwise colonizations, finally covering the whole Galaxy from one end to the other, and every nice planet in it. The complete galactic colonization could well be completed within some 10 million years.

Other scientists, such as Freeman Dyson, Ronald Bracewell, and Eric M. Jones, have reached similar conclusions, with the most conservative time estimate for colonization (that of Jones) set at 300 million years.

With these various futures for ourselves in mind, we can now turn the tables and imagine that any other civilizations that preceded us in this galaxy may have faced the same choices. Two extreme visions mark the limits of the possibilities involved. We have already touched on one of them. Flying saucer advocates maintain that the extraterrestrials have indeed developed in this way and are among us, or at least overhead. Potent evidence would be needed to support such a claim and, as I mentioned, none has appeared. Furthermore, a search using radio telescopes for signals from other star systems has been conducted for some decades, with negative results thus far. In this context, the question "Where are they" comes up. It has been called "the Fermi Paradox," after the Nobel Prize-winning physicist Enrico Fermi, who supposedly raised it for the first time at a Los Alamos, New Mexico, social event in 1943.

Physicist Michael Hart and others have considered this paradox and reached a pessimistic conclusion. The extraterrestrials have not arrived or sent signals to us because they do not exist. Hart is not surprised by their absence, as he has reviewed the same information that we have considered in this book and arrived at the Sour Lemon conclusion. We arose through an exceedingly improbable event and are alone in the universe.

The three-way debate between advocates of Creation, luck, and Cosmic Evolution has been going on for a long time and will not be resolved unless we can collect some new evidence. A quest for evidence of the Life Principle and the search for radio signals from afar represent separate, valid approaches. But we need not limit the strategies we use in tackling a question of such magnitude. A search of likely sites in our solar system for evidence of past visits by extraterrestrials would be a worthwhile approach.

Others who have considered the possibility consider it a longshot. Philosopher Paul Davies has termed it "very unlikely but not impossible," for example. The odds seem against it because it requires that several independent assumptions work out: (1) We are not alone; life has started elsewhere; (2) some of that life has evolved to the point of advanced technology; (3) that civilization has chosen to send robot probes or advance scouting parties to our solar system; and (4) remnants of their visit have survived and can be discovered by us.

This search philosophy requires only that we explore our solar system and find what may be in it. It goes hand in hand with the hunt for existing or evolving life, but it will take us to some sites that are unlikely for the other purpose. It differs from SETI (Search for Extraterrestrial

Intelligence, the radio telescope project that I have mentioned) in an important way.

SETI assumes that the senders are alive now (or were when the signal that they sent started toward us at the speed of light). Flying saucer advocates also assume that the spaceships are operating near Earth right now. Astronomer Michael Papagiannis has suggested that remote probes sent by extraterrestrial intelligence may be orbiting our Sun and returning data. We could locate them by locating a source of tritium (a radioisotope presumed to be produced by a craft's engine) emission in our system. Since this isotope has a life of only a few dozen years, the presumption is made again that the probes are active now.

In searching for the remains of alien artifacts, we only assume that some probe from a distant civilization arrived within the past 4 billion years and left something behind. We are conducting an archeological search that covers a vast time span rather than exploring the present moment in time. We do not have to speculate, initially, as to why it was sent, or what became of the senders. If we find only virginal terrain everywhere that we look within our system, then we will at least have settled one question, and in conducting the search we may achieve another objective. We will show the public that space exploration can focus on the most important questions concerning our existence rather than serve just to collect data of interest to a limited group of planetary specialists.

The risk in such an enterprise is that our search will be confounded with the overimaginative theories of the past, science fiction, and even the contents of supermarket tabloids. The present hunt for life on Mars, for example, is haunted by the canal visions of Percival Lowell, the Martian monsters conjured up by H. G. Wells in the *War of the Worlds*, and the adventures of John Carter. One of my favorite tabloid photos shows a Viking-like scene, allegedly on Mars, with a number of rocks standing erect in a circle in the distance. The caption of the photo carries the claim, "This astonishing photo taken on Mars by *Viking I* lander shows a stone structure uncannily like Stonehenge on Earth," referring to a 4,000-year-old structure on Salisbury Plain in England. The accompanying article contends that federal officials have suppressed this photo for twenty years.

The ultra-conservatism used by NASA representatives in interpreting Gilbert Levin's Viking results may reflect their need to move the agency's image as far as possible from such media hype. But there are better ways to separate serious science from pseudoscience and the occult. To qualify as science, a theory must contain within it the seeds of its own destruction. Critical experiments or observations must be possible that have the potential to prove the idea wrong. Pseudoscience is often marked by the certainty of correctness that accompanies each new idea. For an illustration, consider the famous "Face on Mars."

The *Viking* orbiters photographed a feature in the Cydonia region of Mars that resembles a human face. Richard Hoagland (a science writer who we met in connection with the Europa ocean) and others have argued with some conviction that the object had been constructed by intelligent beings, just as we did for the presidential faces on Mt. Rushmore. Carl Sagan and most scientists concerned with the issue have dismissed it as a natural geological formation, in the same class as the "Old Man of the Mountains" in New Hampshire. Furthermore, they argued, effects of light and shadow as well as black dots from the photographic process were contributing to the effect. The first claim was certainly the extraordinary one, needing extraordinary confirmation. Usually, the burden of proof falls upon those making such a claim. In this case, however, the only way to get more information was through additional photographs by orbiting spacecraft, and the selection of future targets was in the hands of NASA.

For a time the official response of NASA, as documented by Hoagland, was to downplay the issue and avoid committing resources for the survey. Only reluctantly did they include it on their agenda, after a publicity campaign by Hoagland and his supporters that included television appearances, a petition, and picketing outside of Jet Propulsion Laboratory.

Finally, in April 1998, new photographs were obtained by the Mars *Global Surveyor* and released rapidly, without interpretation, on the Internet. No obvious human face was visible. In a report in the *New York Times*, one scientist described the feature as "a butte, a mesa, a knob" and others called it "a sandal print or a chili pepper."

Hoagland was quoted as commenting that the quality of the photographs was too poor to permit any conclusions. Apparently, he chose to ignore one of the key rules of science, that losers and winners emerge when a legitimate test is made. But in this case, I feel that NASA has lost something as well: a glorious chance to educate the American public about the way that science works.

Imagine another course of action, in which NASA invited full media coverage in advance. Interviews, features, or even debates before the event could have highlighted the issue and emphasized the unknown nature of the results that we would obtain. The return of the data might even have earned live television coverage. The public could then have participated in the test of an idea and, in this case, the defeat of a speculation. We would certainly improve upon the current situation, in which science is kept pristinely clear of pseudoscience but the latter gets the larger share of the media and thus flourishes.

If we can move away from our fears of taking up exotic but unlikely ideas, then we may find that one eventually works out. With this spirit in mind, we can start planning a rational search for alien artifacts.

Where Should We Look?

In fishing for remnants of past visitors, we will want to maximize our chances by casting our net as far back in time as possible. Orbiters are likely, over long periods of time, to be deflected or to crash. We should therefore seek landers. The best chance for the preservation of such relics lies on the surface of cold, airless worlds. But we will need to find ones that have been placed in ancient stable terrain, not on planets that have been frequently resurfaced or have undergone other geological upheavals.

To limit our hunt further, we can assume that the landers have been programmed to set down at a site from which they can observe something of interest. In the absence of any other information, I will presume that the ancient visitors were attracted by the same striking features that gain our own attention. Our search can then provide us with another excuse to do what we might want to do anyway: to visit the prime tourist sights of our solar system. In planning an itinerary, I have used a beautifully illustrated book: *The Grand Tour: A Traveler's Guide to the Solar System,* by artists Ron Miller and William K. Hartmann.

For example, consider Mimas, the innermost of Saturn's large moons. A spectacular view of the rings obviously comes with that location. Further, Mimas contains an enormous crater, Hershel by name, that has been called "truly striking." It extends for almost one-third the diameter of the planet, with a broad central peak 4 kilometers high. A site of this type obviously begs for a visit.

There are some drawbacks. According to Miller and Hartmann, "Mimas suffers concentrated bombardment, because it is close to Saturn." Further, the crater has been termed "moderately fresh," so we would not be sampling the full age of our solar system. Yet its intrinsic interest nominates it for any wish list of places to visit.

Miranda, a smallish moon of Neptune, represents another point of interest. Unlike the bland sister moons of this distant planet, Miranda has "bizarre contoured terrain as strange as any surface in the Solar System," according to Miller and Hartmann. They continue: "The single outstanding feature is the most dramatic cliff discovered so far in the Solar System." It rises as much as 5 kilometers over the valley below and has a "breathtaking smooth slope" that descends at about 45 degrees. We can imagine that extraterrestrials might have decided that the cliff was worth a stopover.

Many other sites in our solar system that engage our scientific interest and offer sensational views may have attracted visitors or their probes. If we make the most pessimistic assumptions, then we would be searching for remnants no larger than the *Apollo* and *Pathfinder* spacecrafts, though, of course, larger installations would be welcome. But obviously, a close inspection of the terrain will be needed; a simple fly-by would not do the job. The search will be expensive, and our chances of hitting a jackpot will be slim, so we will not want to design missions for this purpose alone. Our hunt can be tagged onto explorations carried out for other purposes. As the intensive close-range study of the Saturn system, let alone Neptune, seems some time away, we will turn our attention to a target much closer to home.

A Place on the Moon

We will undoubtedly find scenic cliffs and craters when we send robots to explore other solar systems, but the feature that would most claim our attention will be a planet that harbors life. When the *Galileo* spacecraft flew past the Earth in 1990 while en route to Jupiter, it took a variety of measurements on our home world. Carl Sagan and his colleagues analyzed the data as an exercise in life detection, pretending that they were ignorant about the contents of this planet. What did they learn?

They saw continents, oceans, clouds, and the icecaps, and they concluded that liquid water was abundant. They were puzzled by the presence of substantial oxygen in the atmosphere, in particular since detectable amounts of the hydrocarbon methane were also present. Oxygen will eventually react with methane and destroy it, so some source on the planet was continually releasing fresh methane supplies. They also noted that a green substance was widely distributed on the Earth. If that material was involved in photosynthesis, then the presence of oxygen on the planet could be explained. Various lines of indirect evidence pointed to the presence of carbon-based life forms.

The same evidence would have been gathered by any spacecraft that flew by Earth within the past 2 billion years. Before that time, life was present, but without the presence of extensive oxygen in the atmosphere. Yet other clues may have been available that signaled the presence of life to seasoned planetary explorers.

For most of our history, visiting spacecraft would have found our planet a place of unusual interest. If they simply flew by, perhaps after dropping a probe, then no record will remain of their visit. Our search strategy depends on the assumption that they chose, perhaps on a return visit, to establish a stable and secure base for the long-term observation of this planet. If so, one likely place for them to build it would have been on the surface of our Moon. We will therefore want to search that surface for remnants of such an observatory.

Alexey Arkhipov of the Institute of Radio Astronomy in the Ukraine has argued that the surface of the Moon offers a number of advantages for Earth observation, compared to observation from orbit, including stable support and some shielding from radiation and meteorites. If we wanted to get some idea of what we might expect when extraterrestrials visited our Moon, we need only look at our own behavior. We have not been scrupulous about cleaning up.

Science writer Nancy Hathaway has described the litter left by our own astronaut and robot landings on our Moon:

> If visitors from another galaxy ever land on the Moon, they'll know someone else was there first. Millions of years from now, the astronaut's footsteps will still be sunken in the dusty lunar soil, and over twenty tons of huge-ticket, high-tech junk left behind by the American and Soviet space programs will still litter the lunar landscape.

The debris includes crashed satellites, rocket parts, some "barely used moon buggies," seismometers, laser reflectors, tools, equipment, medals, a flag, three cameras, two golf balls, a photograph in plastic of the family of an astronaut, a pin, and a feather from a falcon.

Fortunately, the material is concentrated in a few locations, and most of the Moon's surface has not been disturbed by us as yet. Further, we would hope to recognize the product of an advanced alien technology from one of our own, if the question came up. Arkhipov has pointed out a different problem, however. Geologists who are used to dealing with natural formations might dismiss an artificial object as a natural one. While such an assumption will be correct almost all of the time, we will have no possibility of hitting a jackpot unless we invest some of our energy.

Richard Hoagland's Internet site carries photographs of lunar anomalies bearing names such as "The Lunar Shard" and "The Tower on the Moon." The entire Face on Mars controversy could emerge here again, in a modified form. But scientists again should not just turn their backs for fear of contamination by pseudoscience. The stakes are too high. It would be best if the searchers were aware of the possibilities and had the chance to examine questionable objects at close range.

A successful example was dramatized in Arthur C. Clarke's novel, *2001: A Space Odyssey,* which was based on the screenplay for the film of that name. A satellite reconnaissance of Tycho Crater turns up a magnetic anomaly within it. A human excavation party then uncovers a 3-million-year-old black obelisk, which had been left for our discovery by an advanced race. In that novel, the obelisk was located by the personnel of a permanent and substantial lunar base, as they explored the resources of that world. In reality, as 2001 approaches, no such base exists, as no humans have returned to the Moon for over a quarter century. Fortunately, interest in our neighboring world seems to be reviving, for a number of reasons.

"The Moon Is Back in Style"

This headline in a small December 12, 1997, article in *Science* magazine seemed to represent a change in fortune for that body, which had been neglected in terms of exploration since the last astronauts left. The report anticipated the $63 million Lunar Prospector mission, which departed the next month to map the Moon from an orbit that included the lunar poles. Two months later, a NASA press conference announced that the *Prospector* had indeed struck paydirt. Ice had been detected, mixed with soil, in permanently shaded craters at both lunar poles. The Moon was formed bone dry, but comet infall may have provided a supply of ice over the ages.

Material that landed in sunlight evaporated, but the craters provided a long-lasting refuge.

Ice, of course, is not in short supply on this planet, but on the Moon it would have a greater significance. In the words of Dr. William C. Feldman of Los Alamos National Laboratory, "This is a significant resource that will enable a modest amount of colonization for centuries." Some two thousand people could be maintained for over a century by the amount of water discovered, without recycling, and of course there would be recycling. Further, some water could be split by electrolysis into oxygen and hydrogen, with the former used for breathing and the latter for rocket fuel.

Of course, the benefits will be forgone if no humans return to the Moon to drink the water and breathe the oxygen. NASA's effort sparked a small trend, as various groups, including the European Space Agency, the Japanese, and even some private sources, began to plan future lunar excursions. The plans included orbiters, landers, rovers, and even a sample return, but all to be carried out by robots.

Europeans had hoped to celebrate the millennium with the $342 million EuroMoon project, which would have placed a lander on the rim of Aitkin Crater, at the lunar South Pole. This location has been called "The Most Valuable Real Estate Off Earth" by Alan Wasser, a vice president of the National Space Society. It marks the location of a "strangely lit mountain top standing just above the moon's darkest valley." The summit of that mountain has been dubbed the "Peak of Eternal Light," as its summit is always in sunlight. The Earth can be seen from one side of the mountain but not the other. In practical terms, that location is valuable because it offers uninterrupted solar illumination as a source of power, and a supply of water in the nearby crater.

The European plans had called for a "robotic village" to be set up at the site after the 2001 landing, to carry out scientific studies. Many scientists were unenthusiastic, however, as they considered the mission to be a publicity stunt. Private sources were to supply the bulk of the funding. When this support failed to materialize, the project was called off. It is hoped that another adventure will be planned there for the future. A site that attracts us so much may also have lured earlier visitors.

The LunaCorp of Arlington, Virginia, has considered a different plan for a return visit to the Moon. A Rover would be landed at the *Apollo 11* site and would undertake a 1,000-kilometer trek across the lunar surface. The view would be broadcast back to Earth, and public interest would be stimulated by allowing volunteers (after appropriate training) to take a turn at driving. I don't know what route is planned or activities intended, but I can't think of any goal that would attract more attention as part of the agenda than a search for artifacts.

The search for alien remains would provide an interesting sidelight to future lunar exploration, but it can hardly carry the show, as positive results are unlikely. A number of proposals for commercial development have been made but appear chancy, and routine scientific surveys are unlikely to get substantial funding. If we are to resume our involvement with the Moon, a more coherent and dramatic goal will be needed. I can make a suggestion.

A Lunar Sanctuary

Early in this book, I brought up our need for a new story; one that includes an accurate account of the scientific universe and generates a meaningful human future within it. The philosophy of Cosmic Evolution qualifies. It pictures our existence as the product of a linked series of events that have taken place since the Big Bang. In particular, the events on this planet that produced humans from self-organizing chemicals over the past 4 billion years represent a precious achievement. If we value the experience of human existence, and look forward to its continuation and improvement into the indefinite future, then we do not want to place the accomplishments of 4 billion years of evolution at risk.

But we stand at risk when our biosphere is confined to the surface of our home planet. An asteroid may descend with little warning and wreak havoc with our climate and civilization. Other natural catastrophes that remain unknown to us may have caused some of the earlier extinctions, and they may recur. Finally, there is the threat of nuclear war, which has receded recently but certainly remains as a future possibility. Hopefully, we will escape all of these dangers, but we need an insurance policy.

Imagine that a secure, self-sustaining base existed separate from this planet. It supported a population of humans, together with those advanced species that we have come to depend on. This modern Noah's Ark, unlike its predecessor, would also store the scientific and cultural heritage of our civilization. Computers would preserve most of this bulk, but perhaps some authentic manuscripts and works of art could find their way over as well. If some disaster swept over our home planet, this base could serve as a resource to reseed our civilization on Earth, once the conditions permitted it again.

The Moon would seem to be a logical place for such a base, though Mars or an artificial colony in space may have virtues as alternatives. The closeness of the Moon and its mineral resources give it obvious advantages. Above that, we have already invested our energy and our emotions in that world, and the construction of the base would bring those efforts to some good conclusion. The primary purpose of Moon base would be to ensure human survival, but we could undoubtedly find other good uses for it while awaiting a doomsday that, we hope, never arrives. I am sure that experts will provide a long list of suggestions, including observatories, tourism, and its use as a staging area for interplanetary missions. I hope that the search for alien artifacts has a place on that agenda.

12

Supporting the Dream

In his book *Pale Blue Dot*, the late Carl Sagan wrote about the adventures that awaited us on Mars:

> When I imagine the early human exploration of Mars, it's always a roving vehicle, a little like a jeep, wandering down one of the valley networks, the crew with geological hammers, cameras, and analytical instruments at the ready. They're looking for rocks from ages past, signs of ancient cataclysms, clues to climate change, strange chemistries, fossils or—most exciting and most unlikely—something alive. Their discoveries are televised back to Earth at the speed of light. Snuggled up in bed with the kids, you explore the ancient riverbeds of Mars.

But a page later, other thoughts intruded: "But then I remind myself of the avoidable human suffering on Earth, how a few dollars can save the life of a child dying of dehydration, how many children we could save for the cost of a trip to Mars—and for the moment I change my mind."

There is no rational reason, of course, why these two very different purposes should be competing for resources from the same pot. But even if we narrow our focus and only consider funds that are to be spent on research, similar questions arise. For example, in August 1997 the *Pathfinder* lander (named the Carl Sagan Memorial station) and the Rover *Sojourner* were reporting back data from their position in the Ares Valles region of Mars. They provided details of the weather, the dust, the rock colors and composition, and other routine features of the surface, information of

261

interest to specialists in Martian geology and climatology. Some of their observations confirmed an idea that had already been widely accepted: The area had experienced massive flooding in the early days of Mars. At the end of that mission, a *Nature* summary carried this epitaph: "In short, says one scientist who still praises Pathfinder as an engineering and management experiment, 'our perceptions of Mars haven't changed that much.' " The cost of the Pathfinder mission ran at about $200 million.

During the same month of August, *Nature* also reported that the complete genome sequence of the disease-causing bacterium *Helicobacter pylori* had been determined. This microbe is responsible for stomach ulcers and chronic gastritis and is implicated in stomach cancers. Until about 1984, scientists had assumed that peptic ulcers were caused entirely by stress, diet, and excess gastric acidity. This bacterium may infect as much as half of the world's population, though usually without causing disease.

With the genetic plan of the organism available, scientists will be able to decipher it to devise new strategies for inactivating it. But the interest does not end there. As biochemist Russell F. Doolittle reported in *Nature*, "This is a sequence that has something for everyone—clinicians, sociologists, epidemiologists, biochemists, ecologists, molecular biologists, immunologists, and last mentioned but hardly least interested, evolutionary biologists."

The cost of the project was between 1 and 2 million dollars. The work was performed by by Dr. Craig Ventner and his colleagues at the Institute for Genetic Research in Rockville, Maryland. This private institute bore the cost and made the results freely available to the public. The U.S. National Institutes of Health, of course, is supporting much more massive sequencing efforts, including the entire human genome. Possession of our complete genetic plan will produce enormous benefits for many areas, including medicine, human genetics, anthropology, forensics, and geneology. The entire budget of the National Institute for Human Genome Research for 1997 was under $200 million, less than that spent on Pathfinder. And in May 1998, Dr. Ventner and some colleagues from industry made public a plan by which the entire human genome could be sequenced for that sum. Any competition between Pathfinder and genetic sequencing in terms of the scientific value of the results produced would be declared no contest, in favor of the latter.

What, then, justifies the large expenditures for Martian exploration and other planetary studies? We must look elsewhere than normal science for the answer. The missions deserve support because they may help answer the large cultural questions: How did we get here? Where are we headed?

The missions are important in fashioning a new story for our times, one that incorporates the immense age and size of our universe. But this crucial property may also make them, together with other branches of science that ask global or cosmic questions, the target for bitter attacks. For example, British author Bryan Appleyard has argued that "science is not a neutral or innocent commodity, . . . a convenience. Rather it is spiritually corrosive, burning away ancient authorities and traditions." He considers it inimical to the human spirit and its needs. Appleyard's position apparently influenced the attitude of George Walden, a member of Parliament who in the mid-1980s was the minister in charge of appropriations for civil science. From that position, Walden set up numerous obstacles to scientific spending initiatives, particularly in the area of high-energy physics.

We can also consider the following comment from a *Nature* magazine book review: "James Watt, President Reagan's Interior Secretary . . . explained that we do not need to worry about preserving natural resources because Judgement day is fast approaching. After that we won't need them." In a similar spirit, Creation Science spokesman Duane Gish told me that he felt the public monies spent on space exploration were wasted.

In Western society, the modern scientific picture of a huge, old universe stands in contrast to an earlier one based on the Bible, in which the Earth was the theater for salvation or damnation. It was essential that each of us performs good works to secure our afterlife, but there was no need for us to prepare for an extended human future or to seek ultimate knowledge. Those items were in the hands of God. While the advocates of the earlier system seldom cite it explicitly as a reason for opposing scientific projects, except perhaps in the area of evolution studies, the influence of the earlier worldview is still apparent.

Another group that we met earlier considers human presence in this universe the product of an unlikely accident and expects that mistake to be corrected before too much time has passed. I have termed them the Sour Lemon School. I have cited their views earlier in this book, so I will present just one additional example now, from economist David Pearce:

> Some might argue that humankind has made such a mess of things that it cannot seriously expect to survive much longer, at least in current social forms. Only those who believe in some evolutionary goal for humans may be upset by such a prospect: after all, the end of humankind will not be the end of all species, although many may go down in the process. Earth surely transcends the human race.

It is not as obvious why pessimists should oppose visionary scientific projects as it is for Fundamentalists. Perhaps the pessimists feel that a fly gains dignity if it does not thrash around too much in the web as it awaits the spider. Both motives and others may lie behind the comments of those who see no virtue in the space program. My own mother felt that it was insane to have the *Pathfinder* on Mars: "What are they doing up there?" she asked me. In the same vein, social critic Amitai Etzioni once called the Apollo Program a "moondoggle." I listed a number of comments in the first chapter that presented the same message, but at greater length. Some of these attitudes were anticipated long before the Apollo mission was ever proposed.

At the end of the remarkable 1936 film *Things to Come*, a rocket was set to be launched that would carry the first human beings around the Moon. The launching pad resembled the later Apollo ones remarkably, but H. G. Wells, who wrote the screenplay, was less accurate about the date. He chose the year 2036, not having anticipated that the event would take place as early as 1968, but that is not the point. The rocket missed destruction narrowly as it took off. No technical error was to blame, as in the case of the space shuttle *Challenger*. Rather, a televised agitator had incited a mob by demanding "Who needs progress?" and urged them to destroy the rocket. That outcome was avoided only when the controllers launched the spacecraft ahead of time.

As we saw earlier in this book, negative feelings toward space exploration (though perhaps less violent), coupled with budgetary pressures, have gradually whittled away the planetary program, both manned and unmanned. Apollo gave way to ambitious unmanned interplanetary missions such as the Voyagers, Galileo, and Cassini. These, in turn, have been replaced by less expensive projects such as Pathfinder and NEAR. The public has taken little note of this trend. *Spaceviews* (of the Boston Space Society) reported a 1997 poll in which 85 percent of the respondents "said the government was fairly or very successful in promoting space exploration. . . . However, only a third of the respondents said they strongly supported funding for NASA, compared to about two-thirds for Social Security, Medicare and defense." There is a danger that in the near future the philosophy of "faster, cheaper, better" may be replaced by "cancel now, spend nothing, best of all."

In the face of this trend, I have advocated human exploration of Mars and, after suitable preparation, of Europa, the Saturn system, and points farther out. I coupled this with the idea that we need a permanent self-sufficient base, on the Moon or at another site off this planet. These programs would require Apollo-level expenditures over a period of generations. I tried to justify them as part of an effort to understand the reasons for our existence and ensure our future survival. But they will hardly be cheap. How will the funds be raised? Arguments on this question have always assumed that governments should supply them, as they alone seem to have the necessary resources. But our experience over the past generation has shown that government funding is not reliable enough for projects that extend over long periods of time.

Several examples have come up in this book. In the first chapter, we followed the Apollo project triumphantly through the initial Moon landing and through *Apollo 17*. But the last missions were abruptly terminated. We saw how the delays in the *Galileo* launch may have led to the events that crippled its main antenna. The Cassini mission to Saturn, which is now en route, suffered a close call in 1995 when an appropriations subcommittee of the U.S. Congress voted abruptly to cut off funding for the mission. Fortunately, *Cassini* was saved when the full committee overrode the subcommittee after vigorous lobbying. For the most complete and expensive federal reversal, however, we must move out of space into the area of particle physics (though space may regain the crown if the space station project should finally be abandoned).

The superconducting supercollider (SSC) was a particle accelerator that was recommended in 1984 by a panel of physicists. It was to be the "forefront high-energy facility of the world." After an initial round of competition that involved twenty-five states, the choice narrowed down to Texas versus Illinois. Finally, an area near the town of Waxahachie, south of Dallas, was selected, and construction was begun.

The plans called for an 87-kilometer tunnel 55 meters underground, one big enough to encircle the island of Manhattan. As the tunnel extended, however, the price escalated. The estimate of $4.4 billion became $8.25 billion and then $11 billion. At the same time, criticism from various sources mounted. Author John Lucaks, for example, writing in the *New York Times*, called the SSC "Super Nonsense" and suggested that it may be "one of the greatest boondoggles of all time."

Finally, in October 1993, the project was killed by Congress. Fifteen miles of tunnel had been dug and $2 billion expended. At least another $640 million was needed to satisfy existing contracts, provide separation pay to employees, and seal off the tunnel entrances. George Brown, chairman of the House Science Committee, commented, "It is very difficult for me to perceive why people in their right minds would pursue this course of action."

On that occasion, Stan Wojecicki, chairman of the government's High Energy Physics Advisory Panel, remarked, "This [decision] breaks the very successful partnership between science and government which has extended since the war, starting from the Manhattan Project." He continued, "That partnership involved a level of mutual confidence which has now broken down." But other fields, as we have seen, have suffered similar, if less expensive, reversals. If I want visible proof, I needn't go to Texas. A drive along the north shore of Long Island will bring me to the Shorehaven nuclear reactor, which was built at a cost of over $1 billion and never used. My intent here is not to argue the merits of the SSC or the reactor. But if we desired *not* to have them, there were cheaper ways to do it.

Although blame for each fiasco can probably be assigned to specific individuals and groups, the underlying problem lies within the system. Many groups that form the national electorate are unsympathetic to space exploration because their philosophies include no need to prepare for an extended human future. They may be in the minority when a program is planned, then gain the upper hand at a later time and cancel it. Government in the United States is intrinsically a short-range process, with administrations and congressional majorities safe only until the next election, and appropriations are awarded one year at a time. Small-scale projects and routine scientific undertakings can thrive under such a system, but if we wish to plan projects for the long term, such as a coordinated search for life in the solar system, we will need another way to operate.

These difficulties have not been limited to our type of government or this era. In the century before Columbus, the Chinese emperors of the Ming dynasty launched a global exploration program. Their ships penetrated as far as Java and the eastern coast of the African continent. If the exploration had continued, Chinese ships might have "discovered" Europe, and world history taken a different course. But, according to Mars exploration advocate Robert Zubrin, "the court eunuchs and Confucian bureaucrats who advised the emperor considered the information about the outside world and other civilizations and philosophies to be intrinsically worthless and potentially destabilizing to the divine kingdom, and so they convinced the emperor to have his fleets recalled and the ships destroyed." Zubrin made a direct comparison of this episode to the retreat of the United States from space. The same comparison was made by Smithsonian curator Valerie Neal, who attributed the Ming withdrawal to "a change in the ruling party at home" and a shift of priorities to "domestic concerns: agriculture, roads and the like."

When the Ming rulers discontinued exploration, their place was taken by Europeans, who then left their imprint on the globe. No other government appears ready to replace the United States in its role in massive planetary exploration in the near future, and physicist Richard Gott has made some discouraging projections based on a philosophical idea called the Copernican principle. Gott feels that "there may be only a brief window of opportunity for space travel during which we will in principle have the capability to establish colonies (which could in turn establish further colonies). If we let that opportunity pass without taking advantage of it we will be doomed to remain on the earth where we will eventually go extinct." He suggests that earlier technological civilizations may have also followed this path, which explains why we have not been colonized.

Gott's arguments are based on probabilities and cannot predict the result of our own specific case, particularly if we have the will to do something about it. But if governments, with their massive resources, cannot be relied upon to further such a plan, how then can it proceed?

Many spokespeople have appeared to deplore the decline in federal funding for specific projects or science in general, but they still urge scientists to return and try to extract another crop from the arid terrain. Arthur Clarke's comment has been, "The best advice I can give to the National Space Society and similar organizations is this: despite setbacks and false alarms—continue the search for intelligent life in Washington!" Physicist Robert Park commented in 1996 that the "the total research capacity of the United States is declining for the first time in history." In considering coming budget cuts, he noted morosely that "the cutting will not be left to scientists. It will fall instead to a Congress largely uninformed about science, responding to a public whose naïve overconfidence that science would bring virtually unlimited social benefits has soured in the past few decades into an equally irrational mood of skepticism and distrust." His suggested remedy was that scientists form an effective lobby for basic research.

I have another suggestion: that scientists move to take their future into their own hands. In doing so, they can also act to safeguard the human future. The great churches have shown us how organizations apart from the government can sustain an ideal over the course of generations. Mechanisms of this type could also be used for a secular purpose. I suggest that the search for extraterrestrial life, the colonization of the solar system, and other long-range, large-scale scientific projects of importance to the human future be privately funded—and by a single organization. By operating privately, supporters of these goals would avoid a source of conflict with those who advocate different values.

Many space exploration enthusiasts have assumed the opposite: that we must convince all of our fellows that the enterprise is worthwhile before we can go ahead. For example, Paul Beich led a discussion before the Boston Chapter of the National Space Society in June 1997 and maintained that we needed "a pervasive understanding among all humans of the commonality of our interests—i.e., what is truly good for one of us is good for everyone, and vice versa."

My own personal experience has been different. Duane Gish offered no objection when I suggested that Mars exploration be privately funded. He felt that if anyone was fool enough to spend his own money in that way, he (Gish) wasn't bothered. All manner of elaborate projects are reported in the newspapers, from skyscrapers in Asia to miniature replicas of New York City in Las Vegas, which draw no organized protests. When people don't feel that their *own* tax money is involved, they become much more tolerant of the dreams of others.

Society may be healthier in the long run when groups feel free to follow their own very different philosophies, so long as they stay out of each other's way. If some of us wish to search the planets for life and colonize the stars, while others settle in on Earth to wait for Judgment Day or Doomsday, so be it. Daedalus, a satiric column in *Nature,* may have caught the truth when it jested about the dangers of the opposite course:

All societies believe nonsense. It spreads iresistably through conformity and changes slowly through fashion. As in cell replication, any given society sooner or later hits on a chunk of lethal nonsense, and collapses. In a world of many independent societies, this does not matter. But once communications technology has formed a single global society, the time must come when everyone believes the same lethal nonsense simultaneously. Civilization then collapses globally, leaving no seed from which it could rise again.

A Plethora of Organizations

A number of societies support one or another of the themes I have raised in this book. The National Space Society encourages the various aspects of our space program, while the Planetary Society emphasizes the study of other worlds. The International Society for the Study of the Origin of Life supports meetings and a journal in that field. The First Millenium Foundation targets the project of galactic colonization. Each represents a part of the picture, but as such it has less power than the overall view, which takes in both our origins and our future survival and emphasizes the use of science as our method of choice.

Let me return to a quote from Dr. Lewis Thomas that I presented in Chapter 3: "We talk—some of us, anyway—about the absurdity of the human situation, but we do this because we do not know how we fit in, or what we are for. The stories we used to make up to explain ourselves do not make sense anymore, and we have run out of new stories, at least for the moment."

It is time for a new story. All of the parts may not yet be in place, but it should contain the following features: We will start with an endorsement of science, which has been the foremost tool in our struggle to improve human existence, particularly over the last centuries. Nobel Laureate Leon Lederman has put the point across briefly and well:

> It is science that has converted night to day, extended human longevity, cured many dread diseases, enabled people of very modest means to drive across continents, fly over oceans and surf webs. Following the rules of antiscience (collectively) would condemn the vast majority of humans to extremes of poverty, starvation and early death, allowing the priests and kings to inhabit their drafty castles, monasteries and rectories.

The essence of science is method, not dogma. We listen to nature and absorb its message rather than pasting our preconceptions onto it. The success of this method in improving our material well-being testifies to its validity. Nature's messages, however, have not been limited to medicine and microwaves but have extended to the entire universe. We have learned that it is immensely greater in size and older in age than our ancestors believed a millennium ago.

Science stops at this point. It does not attach value to these findings or create meaning from them. Some of my colleagues have chosen to interpret them in a way that renders humankind insignificant and its existence pointless. But we are the only creatures in sight that seek and create meaning; surely we can find a better message than that. If we find the experience of our lives worthwhile, then we can identify with and value the past processes that have produced it, over the course of billions of years.

Our best information suggests that a unifying principle has governed the development of complexity from the Big Bang to human consciousness, the key feature that governs our awareness. One link in this chain remains disconnected, however. We do not understand how the great leap in organization between nonliving and living systems was bridged. The answer may be found in the planets of our solar system. They were the subjects of myths and dreams for our ancestors; they may hold the solution to one of the great riddles that we face today.

If we value human existence, then we wish to see it extended into the indefinite future that confronts us, one of countless billions of years. That future will allow us time for self-knowledge, to learn to abolish maladies such as disease and aging that most of us would prefer to do without. But most of our future development, our exploration of the human condition, will have to be done somewhere else.

Earth has provided a stable platform for the evolution of life over 4 billion years. But that lease is limited; we know for sure that it will expire after a few billion more. Long before that, our planet may become a place where it is no longer suitable for us to live. Increasing luminosity of the Sun may gradually boil our oceans, or more sudden catastrophes may threaten our existence. If we are wise, we will have furnished our new apartments long before that time. The planets, and then the stars, represent the inevitable path to an extended human future.

The Foundation

In the earliest human societies, stories and traditions were passed by word of mouth from generation to generation. It recent times, books and now computers have stored the collective wisdom of our culture, and appropriate groups have dedicated themselves to its preservation. The viewpoint of Cosmic Evolution will need an organization of its own to further its goals in the coming millennium. I will suggest the name "The Foundation," after the famous science fiction series of Isaac Asimov. His novels were set in the future, when human society had spread through the galaxy, but the name carries a message that applies to our situation today.

Asimov's Foundation was established to secure the future of civilization during a period in which it was threatened by a far-reaching collapse. The organization preserved the knowledge base of the culture and sought to minimize the consequences of any upheaval. It concentrated on political questions, aided by a discipline that allowed it to foresee human history. The Foundation of our own time shares the same purpose—securing the human future—but its focus will cover issues that have a scientific basis.

I have assembled a wish list of tasks for such an organization, but you, the reader, are welcome to add your own ideas.

1. Prepare a coherent plan for the search for life in our solar system. In NASA, at present, missions seem to appear ad hoc, after a competitive process. Past results can be ignored, as in the case of Viking, and several current missions may proceed without reference to one another. An authentic overview would rank the various strategies for finding life and set priorities. The existence of such a plan would not compel NASA or other spacefaring organizations to follow it, but it would focus attention on the question and encourage the development of a coordinated plan.

2. Encourage the construction of authentic simulations of the early Earth, to search for any self-sustaining chemical cycles that might appear. As I explained earlier, most current "prebiotic" syntheses seem more designed to validate the experimenters' preconceptions than to explore openly what might have occurred. Again they represent an effort to force a plan onto nature rather than an effort to learn from it.

3. Evaluate the danger to our civilization from physical processes such as asteroid impacts, global warning, ozone depletion, exhaustion of resources, deep gas eruptions, bursts of gamma rays, and any other threats that may come to light. Decide which of them deserve priority for action, based on the likelihood of the hazard, its impact, and the overall cost.

4. Encourage efforts, such as those in the biosphere projects in Arizona, to design self-sustaining human habitats. Undertake feasibility studies of the relative merits of placing survival bases on the Moon, Mars, and elsewhere. Start the process of assembling databases that preserve the scientific and cultural heritage of our civilization. (In Asimov's novels, this

collection was represented by *The Encyclopedia Galactica*. With computer storage, we can be much more efficient.)

5. Develop programs that encourage public understanding of the universe that we find ourselves in, and the significance of the aforementioned efforts. Some outstanding museum exhibits and television films have been produced, but they exist piecemeal rather than as part of some coherent and integrated presentation. My own nomination would go to the Museum of the Cosmos, which would be good to see in reality rather than just in imagination.

6. Engage in planned long-range fund-raising activity, so that the Foundation can actually carry out some parts of the preceding steps, instead of relying on others. One model for the development phase would be the Howard Hughes Medical Institute, whose assets, according to the Institute's 1997 annual report, were reported to be $10.4 billion. In that year, its disbursements for medical research, special programs, and other activities were over $450 million. A 1997 *Nature* article reported that "the emergence of Hughes over the past decade as a major player in the support of biomedical research means that many of the very best researchers are no longer dependent on public funds, a major departure for biomedical research in the United States."

Hughes is not an isolated example, in terms of its size. In 1994, the *New York Times* listed three foundations with assets over $6 billion, and seven others in the range of $1.5 to $3.7 billion. There is still room for company, as a 1997 editorial written for *Science* pointed out that in the period extending to 2040, "the United States will experience the largest intergenerational transfer of wealth in its history," with the amount likely to be set aside for philanthropic purposes about $1 trillion. But that same editorial noted that "while more than 50% of foundation grant dollars are targeted to education and health and human services, only 4% of grant dollars are targeted to science and technology."

This prioritization has been the result of a number of individual decisions rather that a collective one, as in the federal budget negotiations. Yet both processes have displayed the same deep unwisdom. Over the long run, science and technology have produced much more wealth than they ever have consumed. If anyone wished to assign quantities, they might compare the income, per person, in the United States of 200 years ago with that today. The increase is due to the effects of science and technology. Some industries, such as pharmaceuticals, communications, and computers, are entirely beholden to science for their existence.

To fund the activities of an organization such as the Foundation, and of others dedicated to the advancement of the human future through science, we would only need to divert a small fraction of the wealth produced by technology back into research through some regular mechanism. Perhaps some small fraction of the revenues produced by scientific invention could be returned as royalties to the ultimate sources of that knowledge. In the long run, those expenditures would more than pay for themselves by stimulating new research and inventions. But until we develop such mechanisms, we will have to depend on philanthropy and irregular federal support for our purposes.

Back on the Moon

This book started with Neil Armstrong's well-remembered words from the surface of the Moon. I will close it with a dream in which his successor on that world (or on Mars) expresses the same sentiments, "One giant step for mankind," but carries along with it a commitment to back up that ideal. The following additional words would be very welcome: "This time we stay."

Notes

Chapter 1. Planetary Dreams

p. 1   The quote of John Milton appears in *Paradise Lost*, Book VIII, lines 175–176 (Raphael's advice to Adam). New York: Rinehart & Co., 1953.

p. 2   Details of the Apollo exploration of the Moon can be found in Gene Farmer and Dora Jane Hamblin, *First on the Moon*, Boston: Little, Brown and Co., 1970; Richard S. Lewis, *The Voyages of Apollo*, New York: Quadrangle/The New York Times Book Co., 1974; and in Andrew Chaikin, *A Man on the Moon*, New York: Viking, 1994.

p. 3   The Charles Johnson interview appeared in Robert J. Schadewald, "The Flat-out Truth: Earth Orbits? Moon Landings? A Fraud! Says This Prophet," *Science Digest*, July 1980.

p. 4   Harold J. Morowitz, *Cosmic Joy and Local Pain*, New York: Charles Scribner's, 1987.

p. 8   The hoax published in 1835 in the *New York Sun* has been described in Michael J. Crowe, *The Extraterrestrial Life Debate 1750–1900. The Idea of a Plurality of Worlds from Kant to Lowell*, Cambridge, England: Cambridge University Press, 1986; David S. Evans, "The Great Moon Hoax," *Sky and Telescope*, September 1981, pp. 196–198 and October 1981, pp. 308–310; and Michael J. Crowe, "New Light on the Moon Hoax," *Sky and Telescope*, November 1981, pp. 428–429.

p. 8   The quote from "pseudo-Plutarch" was taken from Steven J. Dick, *The Biological Universe. The Twentieth-Century Extraterrestrial Life Debate and the Limits of Science*, Cambridge, England: Cambridge University Press, 1996.

p. 9   "Especially after Wilkins' book . . ." See Michael J. Crowe, p. 8, op. cit. The quotes from Kepler and Herschel have also been drawn from this source.

p. 11   Johannes Kepler's novel *Somnium* is described in Ron Miller's article, "Astronauts by Gaslight," *Ad Astra*, September–October 1994, pp. 42–45.

p. 13   Donald H. Menzel, "Exploring Our Neighbor World, the Moon," *National Geographic* CXIII, no. 2 (February 1958): 277.

p. 13   Robert Jastrow, "The Exploration of the Moon," *Scientific American* 202 (May 1960): 61–69.

p. 13   The Planetary Society, Pasadena, California, CD: "Visions of Mars," in association with the Time Warner Interactive Group, 1994. Foreword to "The Martians" (1967) by Fred Hoyle. The Hoyle story is reproduced on that CD.

p. 14   Joshua Lederberg and Dean B. Cowie, "Moondust," *Science* 127 (1958): 1473–1475.

p. 14 Harold C. Urey, "The Origin of Some Meteorites from the Moon," *Die Naturwissenschaften* 55 (1968): 49–57.

p. 15 For Edward Anders's offer, see Donald Goldsmith, *The Hunt for Life on Mars*, New York: Penguin Books, 1997, p 110.

p. 16 What we have learned about the Moon is summarized in G. Jeffrey Taylor, "The Scientific Legacy of Apollo," *Scientific American*, July 1994, pp. 40–47, and in Andrew Chaikin, "While We Weren't Watching: Apollo's Scientific Exploration," *Planetary Report* XIV (May/June 1994): 6–9. For a more comprehensive account, see Eric Burgess, *Outpost on Apollo's Moon*, New York: Columbia University Press, 1993. A view immediately after Apollo was presented by Vivien Gornitz and Robert Jastrowin, "Solving the Moon's Riddles," *Natural History*, January 1974, pp. 28–39.

p. 16 The term *Big Whack* and the new estimate of the size of the colliding body are quite recent; earlier calculations had the intruder at about one-tenth of the size of the Earth—a body as massive as the planet Mars; more revisions would not be surprising. See John Noble Wilford in the *New York Times*, September 29, 1997.

p. 18 The political circumstances surrounding Apollo have been described in Andrew Chaikin, *A Man on the Moon*, see p. 2, op. cit., and in Richard S. Lewis, *The Voyages of Apollo*, see p. 2, op. cit.

p. 18 The *Science* magazine with the lunar cartoon on its cover was the issue of August 17, 1973.

p. 19 For Clarke's epilogue, see *First on the Moon*, p. 2, op. cit.

p. 19 Stu Roosa's quote appeared in Chaikin, *A Man on the Moon*, p. 577; see p. 2, op. cit.

p. 20 For the John Noble Wilford quote, "But the wonder," see the *New York Times*, July 17, 1994, p. A1.

p. 20 For the Frank Rich quote, "The notion that Americans . . ." see the *New York Times*, July 5, 1995, p. A21.

pp. 20, 21 The quotes from Richard S. Lewis were taken from *The Voyages of Apollo*, see p. 2, op. cit. Senator Fulbright's remark appeared on p. 171 of this book.

p. 21 The comment by Fred Hoyle was cited by Gustaf Arrhenius and Stephen Mojzsis, "Life on Mars—Then and Now," *Current Biology*, 6, (1996): 1213–1216. The Carl Sagan quote was taken from Carl Sagan, *Pale Blue Dot: A Vision of the Human Future in Space*, New York: Random House, 1994, p. 257. Joseph Corn's comment appeared in "Earthly Woes Supplant Euphoria of Moon Shots," John Tierney, *New York Times*, July 20, 1994, p. A1.

p. 21 Henry Gee's comments were taken from his review of the film *Apollo 13*, published in *Nature* 378, November 2, 1995, p. 103.

p. 22 Bruce Handy, "Fly Me to the Moon," *New York Times Magazine*, July 10, 1994, p. 62.

p. 22 Alex Roland's comment appeared in the *New York Times Book Review*, July 17, 1994, p. 1.

p. 22 Thomas Kuhn, *The Structure of Scientific Revolutions*, 2nd ed., Chicago: University of Chicago Press, 1970.

p. 23 Richard Lewis, *The Voyages of Apollo*, p. 2, op. cit.

p. 25 Kurt Vonnegut Jr., *The Sirens of Titan*, New York: Dell, 1959, p. 7.

p. 25 Paul Davies, *Are We Alone? Philosophical Implications of the Discovery of Extraterrestrial Life*, New York: Basic Books, 1995, p. xii.

Chapter 2. A Shift in the Cosmos

p. 28 Jerome Bruner contrasts the narrative form with logical exposition in *Actual Minds, Possible Worlds*, Cambridge, Mass.: Harvard University Press, 1986.

p. 28 M. Mair, "Stories are habitations . . ." was cited in George S. Howard, "Cultural Tales. A Narrative Approach to Thinking, Cross-Cultural Psychology and Psychotherapy," *American Psychologist*, March 1991, pp. 187–196.

p. 29 Paul C. Vitz's, "Millions of people . . ." appeared in his article, "The Use of Stories in Moral Development. New Psychological Reasons for an Old Educational Method," *American Psychologist*, June 1990, pp. 709–720.

p. 29 T. R. Sarbin's "Our plannings . . ." was quoted in Paul C. Vitz, p. 29, op. cit.

p. 29 The George S. Howard quote, "For example, a child's . . ." appeared in his article; see p. 28, op. cit.

p. 30 My description of the medieval cosmos is derived from the following sources: *The Divine Comedy* of Dante Alighieri, translated by Charles Eliot Norton, Chicago: Encyclopedia Britannica, 1952; Marcelo Gleiser, *The Dancing Universe—From Creation Myths to the Big Bang*, New York: Dutton, 1997; David C. Lindberg, *The Beginnings of Western Science*, Chicago: University of Chicago Press, 1992; and Arthur O. Lovejoy, *The Great Chain of Being*, Cambridge, Mass.: Harvard University Press, 1936.

p. 32 For my account of the last days of the Earth according to Christian theology, I have relied on *Questioning the Millenium*, by Stephen Jay Gould, New York: Harmony Books, 1997, and Gould's article, "The End," in the op-ed section of the *New York Times*, October 23, 1997.

p. 33 William Manchester, *A World Lit Only by Fire*, Boston: Little, Brown and Co., 1992, p. 20.

p. 33 Umberto Eco, *The Name of the Rose*, San Diego: Harcourt Brace Jovanovich, 1980.

p. 34 Connie Willis, *The Doomsday Book*, New York: Bantam, 1992, pp. 137–138.

p. 35 For a discussion of the rules of science, see Robert Shapiro, *Origins: A Skeptic's Guide to the Creation of Life on Earth*, New York: Bantam, 1987, Chapter 1. See also William R. Overton, "Creationism in Schools: The Decision in McLean versus the Arkansas Board of Education," *Science* 215 (1982): 934–943. This article contains the decision of the judge who presided over the Creation Science trial in Little Rock, Arkansas. A longer exposition of the nature of science can be found in Martin Goldstein and Inge Goldstein, *The Experience of Science: An Interdisciplinary Approach*, New York: Plenum, 1984.

p. 38 Kees Boeke, *Cosmic View, the Universe in 40 Jumps*, New York: John Day Company, 1957.

p. 38 Philip and Phyllis Morrison and the office of Charles and Ray Eames, *Powers of Ten*, San Francisco: W. H. Freeman, 1982.

p. 42 Nancy Hathaway, *The Friendly Guide to the Universe*, New York: Penguin Books, 1994, p. 136.

p. 42 Timothy Ferris, *The Whole Shebang*, New York: Simon & Schuster, 1997. This book has served as a valuable source of data in creating the Universe Room.

p. 43 Eric J. Chaisson has also observed that estimates of the age of the universe seem to be converging on 12 billion years. See "Cosmic Age Controversy Is Overstated," *Science* 276 (1997): 1089–1090. However, John Noble Wilford in the *New York Times*,

January 9, 1998, citing Dr. Peter Garnavich of the Harvard-Smithsonian Center for Astrophysics, reported that the universe may be as much as 15 billion years old.

p. 43   The task of mapping a portion of the universe has been described by George Greenstein, "Our Address in the Universe," *Harvard Magazine,* January–February 1994, pp. 39–47. A new mapping effort by Dr. James Gunn of Princeton and his collaborators is now underway. It will map a local slice of the universe in detail to a distance of 1.5 billion light-years. See James Glanz, "What Else Lurks Out There? New Census of the Heavens Aims to Find Out," *New York Times,* March 17, 1998, p. F1.

p. 43   The galaxy count was described by John Noble Wilford in the *New York Times,* January 16, 1996, p. A1.

p. 44   Alan Dressler, *Voyage to the Great Attractor,* New York: Alfred A. Knopf, 1994.

p. 45   The detection of a pair of galaxies so distant that we see them as they were when the universe was 10 percent of its present age was reported by Kenneth Lanzetta, *Nature* 390 (1997): 115–116.

p. 45   "The first low-mass galaxies": Guinevere Kauffman in "The Case of the Missing Ellipticals," *Nature* 390 (1997): 346–347.

p. 45   "Stars were being formed in abundance": For the estimate of the time and rate of peak star formation, see Gretchen Vogel, "Gap in Starbirth Picture Filled," *Science* 276 (1997): 1334.

p. 46   "The new stars then entered into their life cycles": F. Duccio Macchetto and Mark Dickinson, "Galaxies in the Young Universe," *Scientific American,* May 1997, pp. 92–99.

p. 46   Star formation in the Orion Nebula is described by C. Robert O'Dell and Steven V. W. Beckwith, "Young Stars and Their Surroundings," *Science* 276 (1997): 1355–1359.

p. 46   Possible planet formation around MWC480 is described by Vincent Mannings, David W. Koerner, and Annella I. Sargent, "A Rotating Disk of Gas and Dust around a Young Counterpart to β Pictoris," *Nature* 388 (1997): 555–557.

p. 46   "Solar system formation begins . . .": A description of the process of star and planet formation is given by Michael Gaffney, "The Early Solar System," *Origins of Life and Evolution of the Biosphere,* 27 (1997): 185–203.

p. 48   For information on the Akilia Island microfossils, see S. J. Mojzsis, G. Arrhenius, K. D. McKeegan, T. M. Harrison, A. P. Nutman, and C. R. L. Friend, "Evidence for Life on Earth before 3,800 Million Years Ago," *Nature* 384 (1996): 55–59.

p. 48   "How much time was needed for life to begin": See Leslie E. Orgel, "The Origin of Life—How Long Did it Take?," *Origins of Life and Evolution of the Biosphere* 28 (1998): 91–96.

p. 48   Ancient Australian microfossils are described by J. William Schopf, "Microfossils of the Early Archaen Apex Chert: New Evidence of the Antiquity of Life," *Science* 260 (1993): 640–646.

p. 48   The possible inorganic origin of some stromatolytes is discussed by Malcolm Walter, "Old Fossils Could Be Fractal Frauds," *Nature* 383 (1996): 385–424.

p. 50   The Cambrian explosion has been described by Jeffrey S. Levinton, "The Big Bang of Animal Evolution," *Scientific American,* November 1992, pp. 84–91. For more detail see Stephen Jay Gould, *Wonderful Life: The Burgess Shale and Nature of History,* New York: W. W. Norton, 1989.

p. 50   For information on the Chicxulub crater, see H. J. Melosh, "Multi-Ringed Revelation," *Nature* 390 (1997): 439–440.

p. 50   For an account of the late Permian mass extinction, see Douglas H. Erwin, "The Mother of Mass Extinctions," *Scientific American*, July 1996, pp. 72–77, and A. H. Knoll, R. K. Bambach, D. E. Canfield, and J. P. Grotzinger, "Comparative Earth History and Late Permian Mass Extinction," *Science 273* (1996): 452–457.

p. 51   Our Sun's coming collision with a dust cloud was discussed by John Noble Wilford, "Sun's Cruise Through Space Is About to Hit Bumpy Patch," *New York Times*, June 18, 1996, p. C1.

p. 51   For information and the quotes on the death of the Sun, see John Noble Wilford in the *New York Times:* "From Hubble, Dazzling Views of Dying Stars," December 18, 1997, Internet edition, www.nytimes.com; and "The Spectacular Shudders of Dying Stars," *Science Times*, December 23, 1997, p. F1. See also Malcolm W. Browne, "New Look at Apocalypse: Dying Sun Will Boil Seas and Leave Orbiting Cinder," *New York Times*, September 20, 1994, p. C1.

Chapter 3. A Matter of Perspective

p. 52   The quotes of Harold J. Morowitz: are taken from *Cosmic Joy and Local Pain*, New York: Charles Scribner's, 1987, pp. 27–31.

p. 53   The Campanella quote was provided by Steven J. Dick, *Plurality of Worlds. The Origins of the Extraterrestrial Life Debate from Democritus to Kant*, Cambridge, England: Cambridge University Press, 1982, p. 91.

p. 53   Barrie Stavis, *Lamp at Midnight*, Act 1, Scene 5. New York: Bantam Books, 1966.

p. 54   For more information on Galileo, see Nancy Hathaway, *The Friendly Guide to the Universe*, New York: Penguin Books, 1994, pp. 133–147, and Marcelo Gleiser, *The Dancing Universe—From Creation Myths to the Big Bang*, New York: Dutton, 1997, pp. 97–119.

p. 54   The Gallup poll results appeared in Russell Shorto, "Belief by the Numbers," *New York Times Magazine*, December 7, 1997, pp. 60–61.

p. 57   Henry Morris's comments on dinosaurs can be found in "Dragons in Paradise," *Impact*, 241, July 1993, Institute for Creation Research, El Cajun, California.

p. 57   "Are Dinosaurs Alive Today?", *Creation Magazine* 15, no. 4 (September–November 1993): 12–15.

p. 58   The quote from Gish's *Evolution: The Fossils Say No!* appeared in Roger Lewin, "Where Is the Science in Creation Science?", *Science* 215 (1982): 142–146.

p. 59   The quote from Morris and Clark can be found in Overton; see p. 35, op. cit.

p. 59   The quote from the Creation Science Research Center appeared in Ronald L. Numbers, "Creationism in 20th-Century America," *Science* 218 (1982): 538–542.

p. 60   The Gallup poll results were published in "Poll Finds Americans Split on Creation Ideas," *New York Times*, August 22, 1982, p. A22.

p. 60   John Tagliabue, "Pope Bolsters Church's Support for Scientific View of Evolution," *New York Times*, October 25, 1996, p. A1. See also Stephen Jay Gould, "Nonoverlapping Magesteria," *Natural History*, March 1997, p. 16.

p. 61   I have taken the excerpts from *Conversations on the Plurality of Worlds* from two sources:

Lovejoy (p. 30, op. cit.), pp. 130–133, and Sylvia Louise Engdahl, *The Planet-Girded Suns*, New York: Atheneum, 1974, pp. 42–45.

p. 62   The quote from Cardinal Nicholas of Cusa appeared in Steven J. Dick, *The Biological Universe: The Twentieth-Century Extraterrestrial Life Debate and the Limits of Science*, Cambridge, England: Cambridge University Press, 1996, p. 41.

p. 63   "There has been no period . . ." see Lovejoy (p. 30, op. cit.), p. 183.

p. 63   The quotes from Benjamin Franklin and Sir William Herschel were provided in Engdahl (p. 61 op. cit.), pp. 70 and 75.

p. 63   Paul Davies, "The Harmony of the Spheres," *Time*, February 5, 1996, p. 58.

p. 63   The quote from Paine appears on page 166 of Michael J. Crowe, p. 8, op. cit.

p. 64   The first use of the term *scientist* is described by J. H. Fowler in "The Word 'Scientist,' " *Nature*, 114 (1924): 824.

p. 65   William Whewell, *The Plurality of Worlds*, Boston: Gould and Lincoln, 1854.

p. 65   Michael J. Crowe (see p. 8, op. cit.), p. 290.

p. 66   Sir David Brewster, *More Worlds Than One: The Creed of the Philosopher and the Hope of the Christian*, New York: Robert Carter and Brothers, 1854. Some of Brewster's quotes are taken from Crowe (see p. 8, op. cit.), p. 305.

p. 67   Tennyson was cited in Engdahl (p. 30 op. cit.), p. 90.

p. 67   For "the most depressing prediction," see Paul Davies, *The Last Three Minutes*, New York: Basic Books, 1994, p. 9.

p. 67   Information on Hermann von Helmholtz was gathered from the *Encyclopedia Britannica. Macropedia*, vol. 8, Chicago: Encyclopedia Britannica, 1977.

p. 69   The darkening of the galaxies is described by P. J. E. Peebles, D. N. Schramm, E. L. Turner, and Richard G. Kron, "The Evolution of the Universe," *Scientific American*, October 1994, pp. 53–57.

p. 69   The times required for final stellar extinction and further decay of the universe were provided by John Noble Wilford, "At the Other End of the 'Big Bang,' a Possible Whimper," *New York Times*, January 16, 1997, Internet edition, www.nytimes.com. He quoted the extrapolations of Fred Adams and Greg Laughlin, astrophysicists at the University of Michigan.

p. 71   Bertrand Russell, "A Free Man's Worship," in *Why I Am Not a Christian*, a collection of his essays, p. 104. Edited by Paul Edwards and published by Simon & Schuster, New York, 1957.

p. 71   Jacques Monod, *Chance and Necessity*, translated from the French by Austryn Wainhouse, New York: Alfred A. Knopf, 1971.

p. 72   Steven Weinberg, *The First Three Minutes: A Modern View of the Origin of the Universe*, New York: Basic Books, 1977, p. 148.

p. 73   The Leo Tolstoy quote is taken from Carl Sagan, *Pale Blue Dot: A Vision of the Human Future in Space*. New York: Random House, 1994, p. 53.

p. 73   Richard Dawkins, *River out of Eden: A Darwinian View of Life*. New York: Basic Books, 1995, as cited in Gregg Easterbrook, "Science and God: A Warming Trend," *Science* 277 (1997): 890–893.

p. 73   The quote from Christian De Duve is taken from *Vital Dust: Life as a Cosmic Imperative*, New York: Basic Books, 1995, p. xviii.

p. 74 Jacques Monod, see p. 71, op. cit.

p. 75 Lewis Thomas, "On the Uncertainty of Science," *Harvard Magazine,* September–October 1980, pp. 19–22.

p. 76 John Keosian, "Life's Beginnings—Origin or Evolution," in J. Oró et al., eds., *Cosmochemical Evolution and the Origins of Life,* Dordrecht, The Netherlands: Reidel, 1974, pp. 285–293.

p. 77 Eric J. Chaisson, "NASA's New Science Vision," *Science,* 275 (1997): 735. For a more extended discussion of Cosmic Evolution, see Dick (p. 8, op. cit.), pp. 537–554.

p. 77 Information on the Astrobiology Institute and the Gerald Soffen quote can be found in Steve Bunk, "Astrobiology Makes Debut Under NASA," *The Scientist,* June 22, 1998, p. 1.

p. 78 Davies (see p. 25, op. cit.), pp. 21–37.

p. 79 For a well-reasoned account of Intelligent Design, see Michael J. Behe, *Darwin's Black Box: The Biochemical Challenge to Evolution,* New York: Free Press, 1996. See also Philip E. Johnson, *Darwin on Trial,* Washington, D.C.: Regnery Gateway, 1991.

Chapter 4. Life in the Museum

p. 81 For my account of the Paris confrontation, I have relied on John Farley, *The Spontaneous Generation Controversy from Descartes to Oparin,* Baltimore, Md.: Johns Hopkins University Press, 1977, pp. 136–140, and Henry Bastian's reports in *Nature* in 1877: Vol. 15, pp. 314 and 380, and Vol. 16, pp. 276–279. I have also used Farley as a source for much of the following discussion of spontaneous generation. His definition appears on page 1 of his book.

p. 82 The quote from van Helmont is reproduced in D. H. Kenyon and G. Steinman, *Biochemical Predestination,* New York: McGraw-Hill, 1969. Spallanzani's experiment is described on page 17 of this work.

p. 86 Browning's quote is taken from page 75 of Farley (see p. 81, op. cit.).

p. 88 For colorful artwork illustrating how a model of a bacterium might appear at 100,000-fold and 1-million-fold magnification, see Laurence A. Moran et al., *Biochemistry,* Englewood Cliffs, NJ: Prentice Hall, 1994, Figs. 2.2, 2.23, and 2.24.

p. 89 Information on hexokinase has been taken from Donald Voet and Judith G. Voet, *Biochemistry,* 2nd ed., New York: Wiley, 1995, pp. 447–448, and Lubert Stryer, *Biochemistry,* 4th ed., New York: Freeman, 1995, pp. 495, 496, 499. For those who thrive on biochemical detail, I will add a few cautionary notes. The reversal of the heating effect has been shown for many enzymes, but I am not sure whether it works for hexokinase. Also, hexokinase comes apart into two separate strands, but this should not affect my general argument. Further, I am using information gathered on hexokinase isolated from yeast. The bacterial version has been less studied but is likely to be similar.

p. 93 For additional details about ribosomes, consult Peter B. Moore, "The Conformation of Ribosomes and rRNA," *Current Opinion in Structural Biology* 7 (1997): 343–347, and the references therein.

p. 95 The DNA sequence of the bacterium *E. coli* has been published. See Fredrick R.

Blattner et al. (15 co-workers), "The Complete Genome Sequence of *Escherichia coli* K-12," *Science* 277 (1997): 1453–1462.

Chapter 5. The Missing Machine

p. 97 The calculations of Harold Morowitz are presented in his book, *Energy Flow in Biology*, New York: Academic Press, 1968.

p. 99 For a history of the development of genetic concepts, see my book *The Human Blueprint*, New York: St. Martin's Press, 1991.

p. 99 H. J. Muller's article, "The Gene Material as the Initiator and the Organizing Basis of Life," was published in the *American Naturalist*, 100 (1966): 493–517. Carl Sagan's quote was taken from "Radiation and the Origin of the Gene," *Evolution* 11 (1957): 40–55.

p. 99 Richard Dawkins, *The Selfish Gene*, New York: Oxford University Press, 1976.

p. 99–101 For the original experiments of the Spiegelman group, see F. R. Kramer, D. R. Mills, P. E. Coles, T. Nishimura, and S. Spiegelman, "Evolution in Vitro: Sequence and Phenotype of a Mutant RNA Resistant to Ethidium Bromide," *Journal of Molecular Biology* 89 (1974): 719–736, and the references cited within it. For a review of the work of Eigen and his colleagues, see Christof K. Biebricher and William C. Gardner, "Molecular Evolution of RNA in vitro," *Biophysical Chemistry* 66 (1997): 179–192.

p. 101 L. E. Orgel, "Evolution of the Genetic Apparatus," *Journal of Molecular Biology* 38 (1968): 381–393.

p. 101 F. H. C. Crick, "The Origin of the Genetic Code," *Journal of Molecular Biology* 38 (1968): 367–379.

p. 102 Walter Gilbert, "The RNA World," *Nature* 319 (1986): 618.

p. 102 The cited biochemistry text is Lubert Stryer, *Biochemistry*, 4th ed., New York: W. H. Freeman, 1995.

p. 103 Jack Szostak's search for an RNA replicase is reviewed in Alicia J. Hager, Jack D. Pollard Jr., and Jack W. Szostak, "Ribozymes: Aiming at RNA Replication and Protein Synthesis," *Chemistry and Biology* 3 (1996): 717–725.

p. 103 The DNA enzyme is presented in Stephen W. Santoro and Gerald F. Joyce, "A General Purpose RNA-Cleaving DNA Enzyme," *Proceedings of the National Academy of Science USA* 94 (1997): 4262–4266.

p. 103 The quotes from Gerald F. Joyce are taken from his article "The RNA World: Life Before DNA and Protein," pp. 139–151 in *Extraterrestrials: Where Are They?* Ben Zuckerman and Michael H. Hart, eds., 2nd ed., Cambridge, England: Cambridge University Press, 1995.

p. 105 Oparin's 1924 paper, "The Origin of Life," has been translated and reprinted in *Origins of Life: The Central Concepts*, David W. Deamer and Gail R. Fleischaker, eds., Boston: Jones and Bartlett, 1994, pp. 31–71. For a full exposition of Oparin's ideas, see his book *Life, Its Nature, Origin and Development*, translated from the Russian by A. Synge, New York: Academic Press, 1964.

p. 105 The quote from Robert Jastrow was taken from *Until the Sun Dies*, New York: W. W. Norton, 1977, pp. 58–59.

p. 106 For a summary of Miller's studies and related ones, see Stanley L. Miller and Leslie Orgel, *The Origins of Life on the Earth*, Englewood Cliffs, N.J.: Prentice Hall, 1974, and the references therein.

p. 106 Miller's experiment was discussed in *Time* magazine, "Semi-Creation," May 25, 1953, p. 82.

p. 107 The article by George Wald, "The Origin of Life," appeared in the August 1954 *Scientific American*. It has been reprinted in *Life, Origin and Evolution*, with introduction by Clair E. Folsome, San Francisco: W. H. Freeman, 1979.

p. 107 "How many times . . ." The geology text quoted is Richard Foster Flint, *The Earth and Its History*, New York: Norton, 1973, p. 119.

p. 108 The paper by Manfred Eigen and Peter Schuster, "Stages of Emerging Life—Five Principles of Early Organization," was published in the *Journal of Molecular Evolution* 19 (1982): 47.

p. 108 The prebiotic chemistry that I have cited is presented, for the most part, in the book by Miller and Orgel (p. 106, op. cit.).

p. 109 The growth of long RNA chains on mineral surfaces has been described in James P. Ferris, Aubrey R. Hill Jr., Rihe Liu, and Leslie E. Orgel, "Synthesis of Long Prebiotic Oligomers on Mineral Surfaces," *Nature* 381 (1996): 59–61.

p. 109 The quote of James P. Ferris and David A. Usher has been taken from their chapter "Origins of Life" in Geoffrey L. Zubay et al., *Biochemistry*, 2nd ed., New York: Macmillan, 1988, pp. 1120–1151.

p. 109 Robert Shapiro, *Origins: A Skeptic's Guide to the Creation of Life on Earth*, New York: Summit, 1986.

p. 111 For a technical critique of specific failings in prebiotic chemistry, see Robert Shapiro, "The Improbability of Prebiotic Nucleic Acid Synthesis," *Origins of Life* 14 (1984): 565–570; "Prebiotic Ribose Synthesis: A Critical Analysis," *Origins of Life and Evolution of the Biosphere* 18 (1988): 71–85; and "The Prebiotic Role of Adenine: A Critical Analysis," *Origins of Life and Evolution of the Biosphere* 25 (1995): 83–98.

p. 112 Malcolm W. Browne, "Chemist Adds Missing Pieces to Theory on Life's Origins," *New York Times*, July 4, 1996, p. A11.

p. 112 For the cytosine synthesis by Michael P. Robertson and Stanley L. Miller, see "An Efficient Prebiotic Synthesis of Cytosine and Uracil," *Nature* 375 (1996): 772–774.

p. 114 The quote "lagoons are rarely . . ." is taken from R. S. K. Barnes, *Coastal Lagoons*, Cambridge, England: Cambridge University Press, 1980, p. 2.

p. 114 The information presented on lagoons is derived from Barnes (see p. 114, op. cit.); B. W. Logan, *The MacLeod Evaporite Basin, Western Australia*, Tulsa, Okla.: American Association of Petroleum Geologists, 1987; H. D. Holland, *The Chemical Evolution of the Atmosphere and Oceans*, Princeton, N.J.: Princeton University Press, 1984; and B. C. Schreiber, "Introduction," in B. C. Schreiber, ed., *Evaporites and Hydrocarbons*, New York: Columbia University Press, 1988, pp. 1–10.

p. 114 For the data of the Japanese chemists, see S. Sakurai and H. Yanagawa, "Prebiotic Synthesis of Amino Acids from Formaldehyde and Hydroxylamine in a Modified Sea Medium," *Origins of Life* 14 (1984): 171–176.

p. 115 Some of the behavior of cyanoacetaldehyde (CAT) and cyanoacetylene (CECIL) is described in J. P. Ferris, R. A. Sanchez, and L. E. Orgel, "Studies in Prebiotic Synthesis. III. Synthesis of Pyrimidines from Cyanoacetylene and Cyanate," *Journal of Molecular Biology* 33 (1968): 693–704, and in J. P. Ferris, O. S. Zamek, A. M. Altbuch, and H. J. Freeman, "Chemical Evolution XVIII. Synthesis of Pyrimidines from Guanidine and Cyanoacetaldehyde," *Journal of Molecular Evolution* 3 (1974): 301–309.

p. 116   For a picture of the geological configuration needed for RNA synthesis, see G. Arrhenius, B. Sales, S. Mojzsis, and T. Lee, "Entropy and Change in Molecular Evolution—the Case of Phosphate," *Journal of Theoretical Biology* 187 (1997): 503–522.

p. 117   The quote from Cairns-Smith appears in his book, A. G. Cairns-Smith, *Genetic Takeover and the Mineral Origins of Life*, Cambridge, England: Cambridge University Press, 1982, p. 56.

p. 117   For the decomposition of ribose, see Rosa Larralde, Michael P. Robertson, and Stanley L. Miller, "Rates of Decomposition of Ribose and Other Sugars: Implications for Chemical Evolution," *Proceedings of the National Academy of Sciences USA* 92 (1995): 8158–8160.

p. 118   The work of the Eschenmoser group on alternative replicators is summarized in Albert Eschenmoser and M. Volkan Kisakürek, "Chemistry and the Origin of Life," *Helvetica Chimica Acta* 79 (1996): 1249–1259, and Albert Eschenmoser, "Towards a Chemical Etiology of Nucleic Acid Structure," *Origins of Life and Evolution of the Biosphere* 27 (1987): 535–553.

p. 118   Various RNA substitutes are discussed by Alan W. Schwartz, "Speculation on the RNA Precursor Problem," *Journal of Theoretical Biology* 187 (1997): 523–527. For a review of PNA, see Peter E. Nielsen, "Peptide Nucleic Acid (PNA). From DNA Recognition to Antisense and DNA Structure," *Biophysical Chemistry* 68 (1997): 103–108.

p. 118   "Others have suggested that even proteins": Self-replication by peptides (short proteins) is discussed in David H. Lee, Kay Severin, Yohei Yokobayashi, and M. Reza Ghadiri, "Emergence of Symbiosis in Peptide Self-Replication through a Hypercyclic Network," *Nature* 390 (1997): 591–594.

p. 121   A presentation of the clay theory in popular form is presented by A. G. Cairns-Smith, *Seven Clues to the Origin of Life*, Cambridge, England: Cambridge University Press, 1985.

Chapter 6. The Life Principle

p. 123   David Wallechinsky, Irving Wallace, and Amy Wallace, *The Book of Lists*, New York: Morrow, 1977.

p. 125   The quote "Although Teilhard's logic . . ." appears in Monod, *Chance and Necessity* (see p. 71, op. cit.), p. 32.

p. 125   Stephen J. Gould's comments, for the most part, are taken from "War of the Worldviews," *Natural History*, December 1996–January 1997, p. 22.

p. 126   Sir Fred Hoyle, "The World According to Hoyle," *The Sciences* 22 (1982): 9–13.

p. 126   William Jennings Bryan's remarks come from the *Readers Digest*, August 1925, vol. 4, no. 40, reprinted, *Impact*, 213, March 1991, Institute for Creation Research, El Cajun, California.

p. 126 The comments of Henry Morris are taken from *Scientific Creationism* (public school edition), San Diego: CLP Publishers, 1974.

p. 128 James Gleick, *Chaos, Making a New Science*, New York: Penguin Books, 1987, p. 8.

p. 128 Heinz Pagels, *The Dreams of Reason*, New York: Simon & Schuster, 1988, p. 54.

p. 130 The experiments with brass balls were performed by Dr. Paul B. Umbanhowar and his colleagues at the University of Texas and described in the *Scientific American*, November 1996, p. 28.

p. 130 The self-organizing chemical reactions were reported by I. R. Epstein, K. Kustin, P. De Kepper, and M. Orban in "Oscillating Chemical Reactions," *Scientific American*, March 1983, pp. 112–123.

p. 131 Per Bak, *How Nature Works: The Science of Self-Organized Criticality*, New York: Springer-Verlag, 1996. The quote from Al Gore's "Earth in the Balance" appears on page 62 of Bak's book.

p. 131 "Complexity scientists consider . . .": A number of books have dealt with modern complexity theory, including Pagels and Bak (see p. 128 and p. 131, op. cit.). Other examples include M. Mitchell Waldrop, *Complexity: The Emerging Science at the Edge of Order and Chaos*, New York: Simon & Schuster, 1992 and George Johnson, *Fire in the Mind*, New York: Alfred A. Knopf, 1995.

p. 131 Philip W. Anderson, "More Is Different," *Science* 177 (1972): 393–396.

p. 132 John Horgan's remarks are taken from his article, "From Complexity to Perplexity" in *Scientific American*, June 1995, pp. 104–110 and his book, *The End of Science*, Reading, Mass.: Addison-Wesley, 1996, pp. 203–206.

p. 132 Gabriel A. Dover's comments appeared in "On the Edge," *Nature* 165 (1993): 705. This article was a book review of Stuart Kauffman's *The Origins of Order: Self-Organization and Selection in Evolution*, New York: Oxford University Press, 1993.

p. 132 The remark of David Berlinski was published in *Commentary* 102 (September 1996): 34.

p. 133 Stuart Kauffman, *At Home in the Universe*, New York: Oxford University Press, 1995.

p. 134 M. Mitchell Waldrop (see p. 131, op. cit.), p. 128.

p. 135 Some more recent work modeling natural selection on computers has been described by Sandra Blakeslee, "Computer 'Life Form' Mutates in an Evolution Experiment," *New York Times*, November 25, 1997, p. F4.

p. 136 For the ideas of Doron Lancet and collaborators, see Daniel Segré and D. Lancet, "A Statistical Chemistry Approach to the Origin of Life," in *Chemtracts—Biochemistry and Molecular Biology*, Origins of Life Special Issue, Geoffrey Zubay, editor, 465–480 (1999), and D. Segré, Y. Pilpel, S. Rosenwald, and D. Lancet, "Mutual Catalysis in Sets of Prebiotic Organic Molecules: Evolution Through Computer Simulated Chemical Kinetics," *Physica A*, 249 (1998): 558–564.

p. 137 The quote from Günter Wächtershäuser is taken from his article "Order out of Order. Heritage of the Iron-Sulfur World," in J. and K. Trân Thanh Vân, J. C. Mounolou, J. Schneider, and C. McKay, eds., *Frontiers of Life*, Gif-sur-Yvette Cedex, France: Editions Frontières, 1992, pp. 21–39. For a lengthy statement of his theory, see his paper "Groundworks for an Evolutionary Biochemistry: The Iron-Sulfur World," *Progress in Biophysics and Molecular Biology* 58 (1992): 85–201. Recent advances in his theory are discussed by Robert Crabtree, "Where Smokers Rule,"

Science 276 (1997): 222, and Gretchen Vogel, "A Sulfurous Start for Protein Synthesis," *Science* 281 (1998): 627.

p. 139 Paul Davies, "The Harmony of the Spheres," *Time*, February 5, 1996, p. 58.

p. 140 Gerald Feinberg and Robert Shapiro, *Life Beyond Earth: The Intelligent Earthling's Guide to Life in the Universe*, New York: Morrow, 1980.

p. 141 Carl Sagan, *Pale Blue Dot: A Vision of the Human Future in Space*, New York: Random House, 1994, p. 220.

p. 142 The anthropic principle is presented by John D. Barrow and Frank J. Tipler in *The Anthropic Cosmological Principle*, New York: Oxford University Press, 1986.

p. 143 Patrick Glynn, "Beyond the Death of God," *National Review*, May 6, 1996, p. 28.

p. 143 For a discussion of the philosophical consequences of the anthropic principle, including multiple universes and Smolin's natural selection, see John Maynard Smith and Eörs Szathmáry, "On the Likelihood of Habitable Worlds," *Nature* 384 (1996): 107. See also David H. Friedman, "The Mediocre Universe," *Discover*, February 1986, p. 65. Lee Smolin has presented his ideas in a book, *The Life of the Cosmos*, New York: Oxford University Press, 1997.

p. 143 Martin Gardner, "Intelligent Design and Philip Johnson," *Skeptical Inquirer*, November–December 1997, pp 17–20.

Chapter 7. Cosmic Sweepings

p. 144 For more extensive information on asteroids and collisions, see Tom Gehrels, "Collisions with Comets and Asteroids," *Scientific American*, March 1996, p. 54, and Patricia Barnes-Svarney, *Asteroid: Earth Destroyer or New Frontier?*, New York: Plenum, 1996.

p. 144 Timothy Ferris's account, "Is This the End?", appeared in the *New Yorker*, January 27, 1997, pp. 44–55. Paul Davies published his version in *The Last Three Minutes* (see p. 67, op. cit.), pp. 1–7.

p. 145 "The Fifth Planet" is reprinted in Loren Eiseley's *The Star Thrower*, New York: Times Books, 1978, p. 129.

p. 146 Jefferson's comment is posted in the Hall of Meteorites in the American Museum of Natural History in New York City.

p. 147 The quote on the history of the Allende meteorite is taken from Adrian Brearly, "Chondrites and the Solar Nebula," *Science* 278 (1997): 76–77.

p. 147 For my description of the nineteenth-century meteorite quarrel, I have relied on the account of Michael J. Crowe (see p. 8, op. cit.), pp. 400–406.

p. 148 The dispute over the claims of Bartholomew Nagy and his colleagues is described in Steven J. Dick (see p. 8, op. cit.), pp. 369–371.

p. 149 "Many paradoxes are encountered . . ." is from Frank Fitch, Henry P. Schwarcz, and Edward Anders, "'Organized' Elements in Carbonaceous Chondrites," *Nature* 193 (1962): 1123–1125.

p. 149 The evaluation of ancient microfossils was published by J. William Schopf and Malcolm R. Walter, "Archean Microfossils: New Evidence of Ancient Microbes," in J. W. Schopf, ed., *Earth's Earliest Biosphere: Its Origin and Evolution*, Princeton, N.J.: Princeton University Press, 1983, pp. 214–240.

p. 151 For a comprehensive review of meteorite analysis, see John R. Cronin, Sandra Pizzarello, and Dale P. Cruickshank, "Organic Matter in Carbonaceous Chondrites, Planetary Satellites, Asteroids and Comets," in *Meteorites and the Early Solar System*, John F. Kerridge and Mildred Shapely Matthews, eds., Tucson: University of Arizona Press, 1988, pp. 818–857.

p. 151 The amino acid "handedness" in the Murchison meteorite is described in Michael H. Engel and Bartholomew Nagy, "Distribution and Enantiomeric Composition of Amino Acids in the Murchison Meteorite," *Nature* 296 (1982): 837. A summary of the findings concerning amino acids in this meteorite is provided in Christopher F. Chyba, "Left-Handed Solar System?", *Nature* 389 (1997): 234–235. Some possible technical errors in this work have been described by S. Pizzarello and J. R. Cronin, "Alanine Enantiomers in the Murchison Meteorite," *Nature* 394 (1998): 236.

p. 152 The neutron star suggestion was put forward by William A. Bonner and Edward Rubenstein, "Supernovae, Neutron Stars and Biomolecular Chirality," *Biosystems* 20 (1987): 99–111. For the alternative involving dust particles, see Robert Irion, "Did Twisty Starlight Set Stage for Life?", *Science* 281 (1998): 626–627.

p. 152 Mushroom-shaped structures in the Murchison meteorite have been described by Richard B. Hoover, "Meteorites, Microfossils and Exobiology," *Proceedings SPIE* 3115 (1997): 115–136.

p. 153 Information on the ANSMET program can be found in Sara Russell, "Cosmic Messages from the Antarctic," *Ad Astra*, March–April 1997, pp. 24–25. Dr. Roberta Score has published her own account: "The Thrill of the Search: Finding ALH84001," *Ad Astra*, January–February 1997, pp. 5–7.

p. 154 For the evidence concerning ancient Martian life in ALH84001, read David S. McKay, Everett K. Gibson Jr., Kathie L. Thomas-Keptra, Hojatollah Vali, Christopher S. Romanek, Simon J. Clemmett, Xavier D. F. Chillier, Claude R. Maechling, and Richard N. Zare, "Search for Past Life on Mars: Possible Relic Biogenic Activity in Martian Meteorite ALH84001," *Science* 273 (1996): 924–930. A more popular account is presented by David S. McKay, Everett K. Gibson Jr., Kathie Thomas-Keptra, and Christopher S. Romanek, "The Case for Relic Life on Mars," *Scientific American*, December 1997, pp. 58–65.

p. 154 The cited *Nature* editorial appeared in vol. 382, p. 563, August 15, 1996.

p. 156 For criticisms of the argument that ALH84001 holds relics of Martian life, see Donald Goldsmith, *The Hunt for Life on Mars*, New York: Penguin Books, 1997; Edward Anders, "Evaluating the Evidence for Past Life on Mars," *Nature* 274 (1996): 2019–2020; Sharon Begley and Adam Rogers, "War of the Worlds," *Newsweek*, February 10, 1997, p. 56; W. W. Gibbs and C. S. Powell, "Bugs in the Data?" *Scientific American*, October 1996, pp. 20–22; J. P. Bradley, R. P. Harvey, and H. Y. McSween Jr., "No 'Nanofossils' in Martian Meteorite," *Nature* 390 (1997): 454–455; and Jeffrey L. Bada, Daniel P. Glavin, Gene D. McDonald, and Luann Becker, "A Search for Endogenous Amino Acids in Martian Meteorite ALH84001," *Science* 279 (1998): 363–369.

p. 158 For computations and computer simulations of the exchange of meteorites among planets, see B. J. Gladman, J. A. Burns, M. Duncan, P. Lee, and H. F. Levison, "The

Exchange of Impact Ejecta between Terrestrial Planets," *Science* 271 (1996): 1387–1392; and H. J. Melosh, "Mars Meteorite and Panspermian Possibilities," in S. M. Clifford, A. H. Treiman, H. E. Newsom, and J. D. Farmer, eds., *Conference on Early Mars: Geologic and Hydrologic Evolution, Physical and Chemical Environments, and the Implications for Life,* Houston, Tex.: Lunar and Planetary Institute, 1997, pp. 53–54.

p. 158   The survival of bacteria on the Moon has been described in Eric Burgess, *Outpost on Apollo's Moon,* New York: Columbia University Press, 1993, p. 71.

p. 159   The quote from H. J. Melosh (of the Lunar and Planetary Laboratory, University of Arizona) appeared in "The Rocky Road to Panspermia," *Nature* 332 (1988): 687–688.

p. 159   Thomson's remarks were published in *Nature,* August 3, 1871, pp. 262–270.

p. 161   The quote of Hermann von Helmholtz and Hooker's letter to Darwin were provided in Michael J. Crowe, *The Extraterrestrial Life Debate 1750–1900: The Idea of a Plurality of Worlds from Kant to Lowell,* Cambridge, England: Cambridge University Press, 1986, pp. 403–405.

p. 162   The information on radiation-resistant bacteria was reported by Malcolm W. Browne, "Odd Microbe Survives Vast Dose of Radiation," *New York Times,* November 28, 1995, p. C1.

p. 162   "So proponents of Arrhenius's scheme still advocate it . . .": For modern support for radiation-driven panspermia, see Paul Parsons, "Dusting Off Panspermia," *Nature* 383 (1996): 221–222.

p. 164   Information on the Kuiper Belt is presented in Jane X. Luu and David C. Jewitt, "The Kuiper Belt," *Scientific American,* May 1996, pp. 46–52.

p. 165   The quote on medieval theology was reported by John Horgan, "Profile: Fred Hoyle. The Return of the Maverick," *Scientific American,* March 1995, pp. 46–47.

p. 165   "Big-bang cosmology is a form of religious fundamentalism . . ." appears on page 413 of Fred Hoyle's autobiography, *Home Is Where the Wind Blows: Chapters from a Cosmologist's Life,* Mill Valley, Calif.: University Science Books, 1994.

p. 165   For the quasi-steady-state theory, see John Maddox, "The Return of Cosmological Creation," *Nature* 371 (1994): 11.

p. 165   The quote "The breathtaking complexity . . ." is taken from Sir Fred Hoyle, "The World According to Hoyle," *The Sciences* 22 (1982): 9–13.

p. 166   The quote "The overriding intelligence . . ." is from Fred Hoyle, *The Intelligent Universe: A New View of Creation and Evolution,* New York: Holt, Rinehart and Winston, 1983, p. 248.

p. 166   The idea that God equals the universe, and the role of the intelligent silicon chip, is presented in Sir Fred Hoyle and N. C. Wickramasinghe, *Evolution from Space: A Theory of Cosmic Creationism,* New York: Simon & Schuster, 1981, pp. 138–143.

p. 167   The quote "in my days . . ." appears in "The World According to Hoyle" (see p. 165, op. cit.).

p. 167   The presence of bacteria on Venus and other planets is predicted by Fred Hoyle and Chandra Wickramasinghe in *Space Travelers: The Bringers of Life,* Cardiff, England: University College Cardiff Press, 1981, pp. 147–162.

p. 167   The suggestions concerning the arrival of bees and other creatures from space can be found in Fred Hoyle and Chandra Wickramasinghe, *Our Place in the Cosmos*, London: J. M. Dent, 1993. The cause of the Cambrian explosion is discussed on page 571 of the same book.

p. 168 The remarks at the 1877 British Association for the Advancement of Science meeting are reported in Michael Crowe (see p. 8, op. cit.), p. 404.

p. 168 Fred Hoyle's views on epidemics and plagues are summarized in Fred Hoyle and N. C. Wickramasinghe, *Diseases from Space*, New York: Harper & Row, 1979. His views on the AIDS virus were taken from Horgan, "Profile: Fred Hoyle. The Return of the Maverick" (p. 165, op. cit.).

p. 169 John Horgan, p. 165, op. cit.

p. 169 "I myself read in depth . . . ": A more lengthy critique of the ideas and methods of Hoyle and Wickramasinghe can be found in my earlier book, *Origins* (see p. 109, op. cit.), Chapter 9.

p. 170 "Doubtless there will be persons . . ." appears in *Diseases from Space* (see p. 168, op. cit.), p. 140.

p. 170 "For when human beings refuse . . ." is taken from *Our Place in the Cosmos* (see p. 167, op. cit.).

p. 170 "Today we have the extremes" is from Hoyle, *Home Is Where the Wind Blows* (see p. 165, op. cit.), p. 421.

p. 171 For information on the NEAR mission, see Pat Dasch, "The Near-Earth Asteroid Rendezvous Mission," *Ad Astra*, January–February 1996, pp. 32–38; Richard A. Kerr, "Where Do Meteorites Come From? A NEAR Miss May Tell," *Science* 271 (1996): 757; and P. Michel P. Farinella, and Ch. Froeschlé, "The Orbital Evolution of the Asteroid Eros and Implications for Collision with the Earth," *Nature* 280 (1996): 689–691.

p. 171 A comprehensive review of interplanetary and interstellar dust is given by Vladimir A. Basiuk and Rafael Navarro-Gonzalez, "Dust in the Universe: Implications for Terrestrial Prebiotic Chemistry," *Origins of Life and Evolution of the Biosphere* 25 (1995): 457–493.

Chapter 8. A Plentitude of Worlds

p. 172 Stephen J. Gould's quote is taken from his article "A Plea and a Hope for Martian Paleontology," in Valerie Neal, ed., *Where Next Columbus? The Future of Space Exploration*, New York: Oxford University Press, 1994, pp. 107–128.

p. 173 For Stephen J. Pyne's essay "Voyage of Discovery," see Valerie Neal, ed. (p. 172, op. cit.), pp. 9–39.

p. 175 The quote from Harrison Schmitt appears in "A Trip to the Moon," *Where Next Columbus?* Valerie Neal, ed. (p. 172, op. cit.), pp. 41–76.

p. 178 "Parasite Planet" first appeared in *Astounding Stories*, February 1935. It was reprinted in *The Best of Stanley G. Weinbaum*, New York: Ballantine Books, 1974, pp. 75–103.

p. 180 The quotes from Arrhenius and Abbot have been taken from Steven J. Dick (see p. 8, op. cit.), pp. 130–131.

p. 180 For my descriptions of current conditions on Venus, I have relied on the following general sources: David Harry Grinspoon, *Venus Revealed: A New Look below the Clouds of Our Mysterious Twin Planet*, Helix Books, Reading, Mass.: Addison-Wesley, 1997; Carl Sagan, *Pale Blue Dot: A Vision of the Human Future in Space*, New York: Random House, 1994; and Ron Miller and William K. Hartmann, *The Grand Tour: A Traveler's Guide to the Solar System*, New York: Workman, 1993.

p. 182 "Extant life on Venus is out of the question" is taken from L. Colin and J. F. Kasting, "Venus: A Search for Clues to Early Biological Possibilities," in *Exobiology in Solar System Exploration*, G. Carle, D. Schwartz, and J. Huntington, eds., NASA Special Publication 512, Moffett Field, Calif.: NASA, 1992, p. 45.

p. 182 For the quote "it is unlikely that life . . .": See Carl Sagan, "The Planet Venus," *Science* 133 (1961): 849–858.

p. 182 Life in the clouds of Venus is discussed in Feinberg and Shapiro (see p. 140, op. cit.), p. 337.

p. 183 The following references provide a summary of conditions on Jupiter: William B. Hubbard, "Interiors of the Giant Planets," in J. Kelly Beatty and Andrew Chaiken, eds., *The New Solar System*, 3rd ed., Cambridge, Mass.: Sky Publishing Co., 1990, pp. 131–138; Andrew P. Ingersoll, "Atmospheres of the Giant Planets," in Beatty and Chaiken (op. cit.), pp. 139–152; and Peter J. Gierasch, "Dynamics of the Atmosphere of Jupiter," *Endeavor* 20, no. 4 (1996): 144–150.

p. 184 For information about the Jupiter probe, see Richard A. Kerr, "Galileo Hits a Strange Spot on Jupiter," *Science* 271 (1996): 593–594, and "Revised Galileo Data Leave Jupiter Mysteriously Dry," *Science* 272 (1996): 814.

p. 185 The article by Carl Sagan and E. E. Salpeter, "Particles, Environments and Possible Ecologies in the Jovian Atmosphere," appeared in the *Astrophysical Journal*, Suppl. 32 (1976): 737–755. The ideas were discussed and illustrated in Carl Sagan, *Cosmos*, New York: Random House, 1980, pp. 40–43.

p. 186 The quote from R. D. MacElroy appeared in his article, "Life on the Second Sun," in *Chemical Evolution of the Giant Planets*, Cyril Ponnamperuma, ed., New York: Academic Press, 1976.

p. 187 A beautifully illustrated and accessible account of the worlds of the solar system is provided in Miller and Hartmann (p. 180, op. cit.).

Chapter 9. The Big Orange

p. 190 Duane Gish, "The Scientific Case for Creation," in *Evolutionists Confront Creationists*, Frank Awbrey and William W. Thwaites, eds., San Francisco: Pacific Division of the American Association for the Advancement of Science, 1984, pp. 25–37.

p. 190 Norman Horowitz, *To Utopia and Back*, New York: W. H. Freeman, 1986, p. 146.

p. 191 Gilbert Levin's comment appeared in Barry E. DiGregorio, with Gilbert V. Levin and Patricia Ann Straat, *Mars: The Living Planet*, Berkeley, Calif.: Frog, Ltd., 1997, p. 303.

p. 195 Edgar Rice Burroughs, *A Princess of Mars*, New York: Ballantine Books, 1963. This work was originally published in 1912.

p. 195 Ray Bradbury, *The Martian Chronicles*, Garden City, N.Y.: Doubleday, 1950.

p. 195 The quote from the 1907 *Wall Street Journal* was provided by Larry Klaes in his article "The Mars 'Face' and Lowell's 'Canals,' " *Spaceviews*, May 1998, http://www.spaceviews.com/1998/05/

p. 196   The quote is taken from an article by Walter Sullivan, *New York Times*, July 30, 1965, p. 1.

p. 196   The statement about NASA fly-by missions was taken from "Possibility of Intelligent Life Elsewhere in the Universe" (revised October 1977). Report prepared for the Committee on Science and Technology, U.S. House of Representatives, Ninety-Fifth Congress, First Session, Washington, D.C.: U.S. Government Printing Office, 1977, p. 12.

p. 197   An account of the Viking mission can be found in Piers Bizony, *The Rivers of Mars: Searching for the Cosmic Origins of Life*, London: Aurum Press, 1997; and in Henry S. F. Cooper Jr., *The Search for Life on Mars*, New York: Holt, Rinehart and Winston, 1980.

p. 197   The Lewis Thomas quote appeared on the op-ed page of the *New York Times*, July 2, 1978.

p. 198   Michael H. Carr, *Water on Mars*, New York: Oxford University Press, 1996, pp. 57–58. This reference provides a general overview of Martian history and geology.

p. 199   An introduction to the early Mars climate debate can be found in James F. Kasting, "The Early Mars Climate Question Heats Up," *Science* 278 (1997): 1245.

p. 199   The cooler view of conditions on early Mars was presented in Steven W. Squires and James F. Kasting, "Early Mars: How Warm and How Wet?" *Science* 265 (1994): 744–749.

p. 201   The visit of the Pathfinder scientists to the Channeled Scablands is described in "Mars Pathfinder Landing Site Workshop II: Characteristics of the Ares Valles Region and Field Trips in the Channeled Scabland, Washington," LPI Technical Report Number 95-01, Parts 1 and 2, Houston, Tex.: Lunar and Planetary Institute, 1995.

p. 202   Matthew Golombek's and Larry Crumpler's remarks were reported in R. A. Kerr, "Gambling on a Martian Landing Site," *Science* 272 (1995): 347–348.

p. 203   Richard A. Kerr's report, "Pathfinder Tells a Geologic Tale with One Starring Role," appeared in *Science* 279 (1998): 175. For a more complete account of the Pathfinder mission, see Matthew P. Golombek, "The Mars Pathfinder Mission," *Scientific American*, July 1998, pp. 40–49.

p. 203   Madeline Jacobs, "The Wow Reaction," *Chemical & Engineering News*, July 14, 1997, p. 5.

p. 205   Carl Sagan's polar bear comment appeared in the *New York Times*, February 22, 1975, p. A22.

p. 205   The post-Viking quote from Carl Sagan was reported by Henry S. F. Cooper Jr. (see p. 197, op. cit.), p. 121.

p. 205   Roger Rosenblatt, "Visit to a Smaller Planet," *Time*, July 14, 1997, p. 37.

p. 206   The flatness of much of northern Mars was revealed by the Mars Global Surveyor 1996 mission. See Richard A. Kerr, "Surveyor Shows the Flat Face of Mars," *Science* 179 (1998): 1634.

p. 206   Freeman Dyson's suggestion was made in "Warm-Blooded Plants and Freeze-Dried Fish," *Atlantic Monthly* 280 (1997): 71–80.

p. 206   The Antarctic microorganisms within the rocks were described by E. Imre Friedmann in "Endolithic Microorganisms in the Antarctic Cold Desert," *Science* 215 (1982): 1045–1053.

p. 207 The comment by Carl Sagan and Joshua Lederberg appears in *Icarus* 28 (1976): 291.

p. 207 For a technical description of the Labeled Release experiment and Levin's summary of the results, see G. V. Levin, "The Viking Labeled Release Experiment and Life on Mars," *Proceedings SPIE* 3111 (1997): 146–161.

p. 210 My definitions were taken from the *American College Dictionary,* New York: Random House, 1964.

p. 210 An account of the Labeled Release experiment and of Gilbert Levin's post-Viking experiences is given in DiGregorio (p. 191, op. cit.).

p. 211 The report of the Mutch Committee is "A Mars 1984 Mission, Report of the Mars Science Working Group," TM-78419, National Aeronautics and Space Administration, July 1977.

p. 212 The start of the Mars Underground is described in Jon Krakauer's "Descent to Mars," *Air & Space,* October–November 1995, pp. 60–67.

p. 213 Information on the possible history of life on Mars and the way that we might search for it can be found in Christopher P. McKay, "The Search for Life on Mars," *Origins of Life and Evolution of the Biosphere* 27 (1997): 263–289. More detailed discussions are provided in John F. Kerridge, "The Search for Life on Mars," *Proceedings AAAS Annual Meeting,* Seattle, February 13–18, 1997, American Association for the Advancement of Science; and "An Exobiological Strategy for Mars Exploration," NASA SP-530, Exobiology Program Office, NASA, Washington, D.C.: NASA, 1995.

p. 213 For information on earthly underground microbes, see James K. Fredrickson and Tullis C. Onstott, "Microbes Deep inside the Earth," *Scientific American,* October 1996, pp. 68–73. Thomas Gold's ideas on the deep, hot biosphere can be found in "An Unexplored Habitat for Life in the Universe?" *American Scientist* 85 (1997): 408–411.

p. 213 Imre Friedman has reviewed the microorganisms in permafrost: "Permafrost as Microbial Habitat," in D. Gilichinsky, ed., *Viable Microorganisms in Permafrost,* Puschino, Russia: Russian Academy of Sciences, 1994, pp. 21–26.

p. 215 Hot springs on Mars are discussed by Bruce Jakosky in "Warm Havens for Life on Mars," *New Scientist* May 4, 1996, pp. 38–42.

p. 215 "The first, *Mars Surveyor 96,* . . .": The early results from Mars Global Surveyor have been published. See A. L. Albee, F. D. Palluconi, and R. E. Arvidson, "Mars Global Surveyor Mission: Overview and Status," *Science* 229 (1998): 1671–1675 and the accompanying articles.

p. 217 Jack Farmer's comments are included in "Site Selection for Mars Exopaleontology in 2001," on the Internet at http://cmex.arc.nasa.gov/Mars_2001/LSSchedule final.html. Many other articles about this mission are collected at this site. Information about other NASA missions, orbiters, and landers is collected at http://marsweb.jpl.nasa.gov/.

p. 218 "This elaborate quarantine . . . ": The Martian sample quarantine was described briefly by Martin Enserink, "Preventing a Mars Attack," *Science* 279 (1998): 1309.

p. 218 Bruce Jakosky's energy calculations are posted on the Internet in "Exobiological

Considerations for Mars 2001 Landing Sites" at the address given (p. 217, op. cit.) for Jack Farmer.

p. 219 Realistic accounts of human exploration of the surface of Mars have been published as science fiction. See, for example, Ben Bova, *Mars*, New York: Bantam, 1992, and Kim Stanley Robinson, *Red Mars*, New York: Bantam, 1993.

p. 219 For a description of Gusev Crater, see N. A. Cabroll, E. A. Grin, R. Landheim, R. Greeley, R. Kuzmin, and C. P. McKay, "Gusev Crater Paleolake: Two Billion Years of Martian Geologic (and Biologic?) History," posted on the Internet in "Exobiological Considerations for Mars 2001 Landing Sites," at the address given (p. 217, op. cit.) for Jack Farmer.

p. 224 A. G. Cairns-Smith's ideas concerning mineral life have been presented in full technical detail in *Genetic Takeover* . . . (see p. 117, op. cit.). For a more popular account, by the same author, see *Seven Clues to the Origin of Life*, New York: Cambridge University Press, 1985.

p. 224 Gilbert Levin's experiences with bureaucratic bans have been described in DiGregorio (p. 191, op. cit.).

p. 224 Robert Zubrin, *The Case for Mars*, New York: Free Press, 1996.

p. 224 The founding of the Mars Society was reported by Sandra Blakeslee, "Society Organizes to Make a Case for Humans on Mars," *New York Times*, August 18, 1998, p. F3.

p. 224 Future plans for Mars are described by Warren E. Leary, "NASA Still Dreams of Building an Outpost for People on Mars," in the *New York Times*, February 3, 1998, and David Samuels, "Dreams of a Distant Planet," *Civilization*, June–July 1997, pp. 58–66.

Chapter 10. *In the Realm of the Giants*

p. 228 The quotations from Galileo and Marius have been retrieved from the Jet Propulsion Laboratory Galileo Internet site: http://www.jpl.nasa.gov/galileo.

p. 229 The quote of Francesco Silezzi appears in Martin Goldstein and Inge Goldstein, *The Experience of Science: An Interdisciplinary Approach*, New York: Plenum, 1984, p. 295.

p. 229 Biographical material on Galileo has been taken from Lawrence S. Lerner and Edward A. Gosselin, "Galileo and the Spectre of Bruno," *Scientific American*, November 1986, pp. 126–133; the article "Galileo" in the *Encyclopedia Britannica*, 15th ed., *Macropedia*, vol. 7, pp. 851–853, Chicago: Encyclopedia Britannica, 1978; and Michael J. Crowe, *Theories of the World from Antiquity to the Copernican Revolution*, New York: Dover, 1990.

p. 229 For a fine summary of our knowledge prior to the Galileo mission, see Ron Miller and William K. Hartmann (p. 180, op. cit.).

p. 229 An account of the Galileo mission is given by Torrence V. Johnson in "The Galileo Mission," *Scientific American*, December 1995, pp. 44–52.

p. 230 For an overview of current information on the moons of Jupiter, see William B. McKinnon, "Galileo at Jupiter—Meetings with Remarkable Moons," *Nature* 390 (1997): 23–26. This picture, of course, is subject to revision. For information about the possibility of an underground ocean on Callisto, see Richard A. Kerr, "Geophysicists Ponder Hints of Otherworldly Water," *Science* 279 (1998): 30–31; and J. D. Anderson, G. Schubert, R. A. Jacobson, E. L. Lau, W. B. Moore, and W. L. Songren, "Distribution of Rock, Metals, and Ices in Callisto," *Science* 280 (1998): 1573–1575.

p. 231   The comment by Margeret Kivelson was reported in Richard A. Kerr, "An Ocean for Old Callisto," *Science* 280 (1998): 1695.

p. 232   The proceedings of the 1979 conference in which Gerald Feinberg discussed our ideas were subsequently published: Robert Shapiro and Gerald Feinberg, "Possible Forms of Life in Environments Very Different from the Earth," in *Extraterrestrials: Where Are They?*, Michael H. Hart and Ben Zuckerman, eds., New York: Pergamon Press, 1982, pp. 113–121. For a longer discussion, see Feinberg and Shapiro, *Life Beyond Earth* (p. 140, op. cit.), pp. 328–332. Ralph Greenberg, a mathematics professor at the University of Washington, has traced the development of the idea of undersea life on the Jovian moons through a number of sources in the 1970s.

p. 232   For information about life in deep sea hot vents, see Nils G. Holm, "Why Are Hydrothermal Systems Proposed as Plausible Environments for the Origin of Life?" *Origins of Life and Evolution of the Biosphere* 22 (1992): 5–14.

p. 232   Captain Cook's quote appeared in Robin Hanbury-Tenison, *The Oxford Book of Exploration*, Oxford, England: Oxford University Press, 1994, p. 492.

p. 233   The March 1998 press release on Europa was reported in the *Boston Globe*, "Water Is Sighted on a Jupiter Moon," March 3, 1998, p. A08. See also the JPL Internet site (p. 228, op. cit.).

p. 234   Arthur C. Clarke, *2010: Odyssey Two*, New York: Ballantine Books, 1982.

p. 235   Richard Hoagland's January 1980 *Star & Sky* magazine article has been posted at his Internet site: www.enterprisemission.com/europa.htlm.

p. 235   The abstracts of the Europa Ocean Conference, November 12–14, 1996, were published by the San Juan Capistrano Research Institute, San Juan Capistrano, California. See the Internet site http://www.sji.org/conf/eurconf/form.html.

p. 236   The evidence that Europa's ocean is salty has been reviewed by Jeffrey S. Kargel, "The Salt of Europa," *Science* 280 (1998): 1211–1212.

p. 237   The possibility of freeze-dried fish in Jupiter orbit was suggested by Freeman J. Dyson, "Warm-Blooded Plants and Freeze-Dried Fish," *Atlantic Monthly*, November 1997, pp. 71–80. Henry Harris of Jet Propulsion Laboratory has suggested an Ice Clipper mission in which we would hit Europa with a 10-kilogram (22-pound) impactor, capture some of the debris in space, and return it to Earth. We would be unlikely to collect fish from this small tap, but other evidence of life might be collected. See the abstracts of the 1996 San Juan Capistrano conference on Europa's ocean (p. 235, op. cit.).

p. 237   A description of a proposed Europa orbiter mission has been posted on the Internet at http://www.jpl.nasa.gov/ice_fire/europao.htm. See also B. C. Edwards et al., "The Europa Ocean Discovery Mission," *Proceedings SPIE* 3111 (1997): 249–261.

p. 237   The "cryobot"-"hydrobot" plan for Europan exploration was described by Warren E. Leary, "Hardier Breed of Antarctic and Lunar Explorers: Robots," *New York Times*, May 13, 1997, p. C4.

p. 237 For information on Lake Vostok, see Richard Stone, "Russian Outpost Readies for Otherworldly Quest," *Science* 279 (1998): 650–651.

p. 239 Kurt Vonnegut Jr., *The Sirens of Titan*, New York: Dell, 1959, p. 265.

p. 239 For information on the ocean and other features of Titan, see Jonathan J. Lunine, "Does Titan Have Oceans?" *American Scientist*, March–April 1994, pp. 134–143; and Stanley F. Dermott and Carl Sagan, "Tidal Effects of Disconnected Hydrocarbon Seas on Titan," *Nature* 374 (1995): 238–240.

p. 240 Jonathan Lunine of the University of Arizona has also proposed a model in which Titan's ocean lies underground, below a porous crust. See "Does Titan Have Oceans?" p. 239, op. cit.

p. 240 The NASA poster described was furnished by the Jet Propulsion Laboratory, Pasadena, California, JPL 400–576 6/96. Craig Attebery is the artist.

p. 240 The article that described the Cassini mission was Elizabeth K. Wilson, "Liftoff to Saturn," *Chemical and Engineering News*, October 13, 1997, pp. 25–27.

p. 242 The quote of Toby Owen and his colleagues appeared in T. Owen, D. Gautier, F. Raulin, and T. Scattergood, "Titan," in *Exobiology in Solar System Exploration*, G. Carle, D. Schwartz, and J. Huntington, eds., NASA Special Publication 512, Moffett Field, Calif.: NASA, Ames Research Center, 1988, pp. 127–143.

p. 242 The quote concerning life's water requirement appeared in David W. Clarke and James P. Ferris, "Chemical Evolution on Titan: Comparisons to the Prebiotic Earth," *Origins of Life and Evolution of the Biosphere* 27 (1997): 225–248. This article contains a detailed review of the possible chemical processes in Titan's atmosphere.

p. 243 For Carl Sagan's ideas concerning temporary lakes of water on Titan, see *Pale Blue Dot* (p. 141, op. cit.), p. 111.

p. 244 The Cassini-Huygens mission has been described by Jonathan I. Lunine and Ralph D. Lorenz in "The Surface of Titan Revealed by Cassini-Huygens," *Proceedings of SPIE—The International Society for Optical Engineering* 2803 (1996): 45–54.

Chapter 11. Signs of Ancient Visitors

p. 245 Planetfest 97 was described by Jennifer Vaughn in "Mars Pathfinder Landing Inspires Thousands Celebrating the Adventure of Exploration," *Planetary Report* XVII, no. 6 (November–December 1997): 4–9.

p. 246 The *Time* magazine article on aliens is Bruce Handy, "Roswell or Bust," *Time*, June 23, 1997, pp. 62–71. For further Roswell-related material, see William Sims Bainbridge, "Extraterrestrial Tales," *Science* 279 (1998): 671; a book review of Benson Saler, Charles A. Ziegler, and Charles B. Moore, "UFO Crash at Roswell," Smithsonian Institute Press, Washington, D.C., 1998; Kal A. Korff, "What *Really* Happened at Roswell," *Skeptical Inquirer*, July–August 1997, pp. 24–31; and Bernard D. Gildenberg and David E. Thomas, "Case Closed: Reflections on the 1997 Air Force Roswell Report," *Skeptical Inquirer*, May–June 1998, pp. 31–36.

p. 246 The account of the technician who experienced an abduction can be found in William H. Honan, "Harvard Investigates Professor Who Wrote of Space Aliens," *New York Times*, May 4, 1995, p. A18.

p. 247 For information on Project Mogul, see Dave Thomas, "The Roswell Incident and Project Mogul," *Skeptical Inquirer,* July–August 1995, pp. 15–18, or access the U.S. government report, "The Roswell Report: Case Closed," at http://www.access.gpo.gov/index.html.

p. 250 A brief description of the 1994 interstellar probe meeting has been published: Larry Krunemaker, "Visionaries Swap Pointers on Star Flight," *Science* 286 (1994): 212–213. The 1998 meeting was described by James Glanz, "Engineers Dream of Practical Star Flight," *Science* 281 (1998): 765–766.

p. 250 The reference *Extraterrestrials: Where Are They?*, Ben Zuckerman and Michael H. Hart, eds., 2nd ed., Cambridge, England: Cambridge University Press, 1995, is a good source of information on the possible colonization of the galaxy. See in particular the articles by Sebastian van Hoerner, "The Likelihood of Interstellar Colonization, and the Absence of Its Evidence," pp. 29–33, Ronald Bracewell, "Pre-emption of the Galaxy by the First Advanced Civilization," pp. 34–39, Ian Crawford, "Interstellar Travel: A Review," pp. 50–69, and Eric M. Jones, "Estimates of Expansion Timetables," pp. 92–102. Freeman Dyson has written on these same issues in *Disturbing the Universe,* New York: Harper & Row, 1979.

p. 251 Michael Hart's ideas are presented in two articles in the book *Extraterrestrials: Where Are They?* (p. 250, op. cit.): "An Explanation for the Absence of Extraterrestrials on Earth," pp. 1–8; and "Atmospheric Evolution, the Drake Equation and DNA: Sparse Life in an Infinite Universe," pp. 215–225.

p. 251 Paul Davies, *Are We Alone?* (see p. 25, op. cit.), p. 40.

p. 252 Michael Papagiannis's suggestion about tritium sources, "A Search for Tritium Sources in Our Solar System May Reveal the Presence of Space Probes from Other Stellar Systems," appears in the book *Extraterrestrials: Where Are They?* (see p. 250, op. cit.), pp. 103–107.

p. 252 For Stonehenge on Mars, see Mike Foster, "Space Probe Takes Amazing Photo of Stonehenge," *Weekly World News,* June 24, 1997, pp. 1–3.

p. 253 Richard C. Hoagland's views concerning the face on Mars are described in his book *The Monuments of Mars: A City on the Edge of Forever,* 4th ed., Berkeley, Calif.: North Atlantic Books, 1996. See also his Internet page at http:/www.enterprisemission.com/bridge.html.

p. 253 For Carl Sagan's views on the face on Mars and claims of extraterrestrial visitation, see *The Demon Haunted World,* New York: Random House, 1995.

p. 253 The *New York Times* report on the face on Mars, "New Mars Photos Cast Doubt on Speculation on a 'Face,'" appeared on April 7, 1998, p. A24.

p. 254 The descriptions of Mimas and Miranda have been taken from Miller and Hartmann (see p. 180, op. cit.), pp. 161–166.

p. 255 For the analysis of the *Galileo* Earth fly-by data, see Carl Sagan, W. Reid Thompson, Robert Carlson, Donald Gurnett, and Charles Hord, "A Search for Life on Earth from the Galileo Spacecraft," *Nature* 365 (1993): 715–721.

p. 256 For Arkhipov's reasoning, see Alexey V. Arkhipov, "A Search for Alien Artifacts on the Moon," in G. Seth Shostak, ed., *Progress in the Search for Extraterrestrial Life,* ASP Conference Series, 74, 1995, pp. 259–264.

p. 256 Nancy Hathaway, *The Friendly Guide to the Universe*, New York: Penguin Books, 1994, p. 39.

p. 257 For information on the discovery of ice on the Moon by the *Lunar Prospector*, see Warren E. Leary, "Craft Sees Signs of Water as Ice in Moon Craters," *New York Times*, March 6, 1998, p. A1, and "Lunar Prospector Finds Water on the Moon," *Spaceviews Update 1998*, March 15 (Boston Chapter, National Space Society), available on the Internet at http://www.spaceviews.com/1998/0315/.

p. 258 Alan Wasser's article "The Most Valuable Real Estate off Earth" appeared in *Ad Astra*, May–June 1995, p. 29.

p. 258 Plans for future lunar exploration can be found in *Spaceviews Update 1998*, March 15 (p. 257, op. cit.); and Tony Reichhardt, "Ice on the Moon Boosts Hopes for Future Lunar Missions," *Nature* 392 (1998): 111.

Chapter 12. Supporting the Dream

p. 261 Carl Sagan's remarks are taken from *Pale Blue Dot* (p. 141, op. cit.), pp. 261 and 263.

p. 262 The quote from *Nature* on Pathfinder came from Tony Reichhardt, "Does Low Cost Mean Low-Value Missions?", *Nature* 389 (1997): 399.

p. 262 A summary of the bacterial sequencing work has been provided by Russell F. Doolittle, "A Bug with Excess Gastric Avidity," *Nature* 388 (1997): 515–516. The full details have been published: Jean-F. Tomb et al. (forty-one authors), "The Complete Genome Sequence of the Gastric Pathogen *Helicobacter pylori*," *Nature* 388 (1997): 539–547. See also Nicholas Wade, "Scientists Map a Bacterium's Genetic Code," *New York Times*, August 7, 1997.

p. 262 The plan to sequence the human genome for $200 million was described by Nicholas Wade: "Scientist's Plan: Map All DNA Within 3 Years," *New York Times*, May 10, 1998, p. A1.

p. 263 The James Watt comment comes from "Is God Green," a book review by John Adams of Michael Zimmerman's *Science, Nonscience and Nonsense, Approaching Environmental Literacy*, *Nature* 381 (1996): 125.

p. 263 The quote from David Pearce appears in "Money Matters," *Nature* 380 (1996): 295–296.

p. 264 For the poll, see *Spaceviews*, April 1997, which is available on the Internet at http://www.seds.org/spaceviews/9704/.

p. 265 Cassini's close call was described by Tony Reichhardt, "Congress Saves Cassini, But Targets Infrared Astronomy Mission," *Nature* 376 (1993): 284.

p. 265 For information on the SSC, see "SSC Decision Ends Postwar Era of Science-Government Partnership," *Nature* 365 (1993): 773, and Gary Taubes, "The Super-collider: How Big Science Lost Favor and Fell," *New York Times*, October 26, 1993, p. C1. The article by John Lukacs, "Atom Smasher Is Super Nonsense," appeared in the *New York Times*, June 17, 1993, p. A25.

p. 266 Robert Zubrin's comments, "The Universe Presents Its Challenge," appeared in *Ad Astra*, March–April 1997, p. 2.

p. 267 See Valerie Neal, *Where Next Columbus?* (p. 172, op. cit.), pp. 195–221.

p. 267   See J. Richard Gott III, "Implications of the Copernican Principle for Our Future Prospects," *Nature* 363 (1993): 315–319, and "A Grim Reckoning," *New Scientist*, November 15, 1997, pp. 36–39.

p. 267   Arthur Clarke's remark appeared in "When Will the Real Space Age Begin?" *Ad Astra*, May–June 1996, pp. 13–15.

p. 267   Robert L. Park, "Fall From Grace," *The Sciences*, May–June 1996, pp. 18–21.

p. 268   Paul Beich, "Ruminations of a Space Activist," *Spaceviews*, July 1997, available on the Internet at http://www.seds.org/spaceviews/9707/.

p. 268   The cited Daedalus article, "Save the Earth," appeared in *Nature* 361 (1993): 408.

p. 269   Leon M. Lederman, "A Strategy for Saving Science," *Skeptical Inquirer*, November–December 1994, pp. 23–28.

p. 271   The first novel in Asimov's series was *Foundation*, New York: Doubleday, 1951, reprinted by Bantam, 1991.

p. 272   For the information on the Howard Hughes Medical Institute, see Colin Macilwain, "Hughes Confirms Its Faith in Excellence," *Nature* 387 (1997): 223.

p. 272   The assets of foundations were compiled in a table: "Top Foundations" in the *New York Times*, May 6, 1994, p. B9.

p. 272   The projections on private philanthropy were taken from Susan M. Fitzpatrick and John T. Bruer, "Science Funding and Private Philanthropy," *Science* 277 (1997): 621.

Index